A Colour Atlas of
Hand Conditions

W. Bruce Conolly

FRCS, FRACS, FACS
Honorary Surgeon to Sydney Hospital, St Luke's Hospital, and
Surgeon in Charge, Sydney Hospital Hand Clinic,
Sydney, Australia

Wolfe Medical Publications Ltd

Copyright © W. B. Conolly, 1980
Published by Wolfe Medical Publications Ltd, 1980
Printed by Smeets-Weert, Holland
ISBN 0 7234 0739 8

This book is one of the titles in the series of
Wolfe Medical Atlases, a series which brings
together probably the world's largest systematic
published collection of diagnostic colour
photographs.
For a full list of Atlases in the series, plus
forthcoming titles and details of our surgical,
dental and veterinary Atlases, please write to
Wolfe Medical Publications Ltd, Wolfe House,
3 Conway Street, London W1P 6HE.

Series Editor: G. Barry Carruthers, MD(Lond)

Contents

Acknowledgements

I am deeply indebted to Mr J.P.W. Varian for his help and advice, and for providing many of the photographs.

I am grateful to Mr H.B. Kapila and Drs L.G. Abbott and H.F. Molloy for their contribution to chapters 10 and 11.

I also express appreciation to other friends and colleagues who lent photographs, especially to: Grace Warren, A. Pelly, and D. Tracy; and also N. Barton, P. Brown, R.E. Carroll, D.K. Faithfull, F. Harvey, J.N. Lavan, W. McBride, W. McCarthy, J. McCredie, J. McCleod, G.W. Milton, A. Mitchell, J.W. Niesche, S. O'Riain, G. Pulvertaft and D. Seaton.

The following authors of other atlases in this series were kind enough to allow the use of several of their pictures: Dr A.C. Boyle, Professor G.A. Gresham, Mr R.T. Hutchings, Dr G.M. Levene, Professor R.M.H. McMinn, Mr W.F. Walker, Dr A. Wisdom, and Dr M. Zatouroff.

Edward Arnold Ltd and the Journal of the Irish Medical Association gave permission to reproduce illustrations.

The atlas could not have been compiled without the generous help of Mrs Anne Hutchings and Mr John Collins for the drawings, Mr Reginald Money for photographic assistance, Miss Margaret Power for help in the library and Miss Eveline Gallard for preparing the manuscript and index.

This book is dedicated to
Joyce, John, Christine, and Bruce
and to all the patients

Preface

Man depends on his hands for his work, recreation and expression. The successful treatment of any hand disorder will depend on an accurate initial diagnosis, and it is towards that end this atlas has been produced.

No organ, anatomical structure, or laboratory procedure can reveal as much practical information about a patient as can the hand. Nor is there any other part more convenient or accessible for clinical examination.

This colour atlas has been produced to assist in the diagnosis of hand conditions. Normal hand anatomy and function are given substantial coverage as an understanding of these is necessary before disorders can be appreciated. The bulk of the atlas comprises a selection of clinical photographs, accompanied by brief captions and, wherever it is useful, by operative pictures, x-rays, diagrams, tables and charts. The appendices include various glossaries and aids to differential diagnoses. The atlas is a companion to, and not a substitute for, a textbook of the hand. For more information about the nature and treatment of each condition the reader should refer to one of the standard textbooks.

Medical students and candidates for undergraduate and post-graduate examinations, both medical and surgical, will find the atlas a useful adjunct to their studies. Those in medical and paramedical practice whose patients present either directly or indirectly with a hand disorder should also find useful information – physicians, dermatologists, rheumatologists, rehabilitation physicians, paediatricians, surgeons (general, plastic, orthopaedic, reconstructive and paediatric), casualty doctors and emergency physicians, nurses, physiotherapists, occupational therapists and splintmakers.

Although there are photographs of patients and conditions from many different countries, the atlas cannot be comprehensive. To include every hand condition would require several volumes. Most of the clinical photographs were taken by the author in outpatient clinics and during operations in the USA, UK, and Australia. Where the photographs have been taken during operation the skin may appear slightly yellowish because of the skin preparation solution used to clean it.

Introduction

There are two questions to ask in diagnosis of a hand condition:

(1) *What is the tissue or tissues of origin of the condition?* (Does it arise from the skin, palmar fascia, flexor tendon etc?)

(2) *What is the type of pathology affecting that tissue?* (Is it congenital, traumatic, infective, neoplastic etc?)

To answer these questions follow the traditional sequence of history, examination, and investigation.

History

Ask about the time and nature of the onset and progress of the disorder. This may provide sufficient information for the diagnosis. For example, burning pain in the hand, radiating to the shoulder and waking the patient from sleep is probably due to a carpal tunnel syndrome.

Examination

Look, feel and test the function of the hand and each of its tissues.

(a) Inspection. Look first at the posture of the hand. Look at each surface: palmar, dorsal, ulnar, and radial. Swelling is always maximal on the dorsum of the hand. Deformities, e.g. finger contractures, are best seen from the side. Look at both hands – arthritis is usually bilateral. Look at the whole patient – xanthomas occur on the hand and eyelids; spider naevi are associated with liver enlargement.

(b) Palpation. Both hands and forearms of the patient should be bare and supported. Use your finger pulp to feel for texture, the dorsum of your finger to test for sweating and temperature. Use a blunt probe to find local tenderness.

Test for fluctuation, but note that the normal pulp is fluctuant.

(c) Test function. Movement, power, sensation, and circulation. See Chapter 1.

Test any swelling for translucency – ganglia and lipomas are translucent.

X-ray

An x-ray is invaluable in the diagnosis of hand disorders. Three views may be required, e.g. to localise a foreign body.

X-rays outline soft tissues as well as the skeleton. They may also show calcification.

The following three tables show the routine for an examination of the hand.

Routine for examination of the hand

1. General examination of both hands, both upper limbs, and the head and neck

REGION	SIGN	SIGNIFICANCE
Both hands	Abnormal posture, deformity	Congenital deformity Dupuytren's contracture, arthritis Skeletal, tendon or nerve injury Hysteria. Nerve disorders
	Scars, sinuses	Injury, infection, previous operation
	Tremor	Functional or organic (Parkinson's disease)
	Swelling	Generalised: injury, infection, lymphovascular disorder Localised: tumour, arthritis
	Range of movements, motor or sensory loss	
Both upper Limbs	Scar, deformity	Injury to deep structures
	Joint swelling, deformity	Arthritis
	Motor or sensory loss	Brachial plexus palsy
	Muscle tone and wasting	
	Range of movements – shoulder, elbow, wrist	
	Reflexes	

Head and neck	Tenderness over posterior neck
	   muscles
		Cervical spondylosis with or without nerve
		   root compression (brachial neuralgia)

	Tinel's sign over brachial plexus, restricted cervical movements	Cervical rib, thoracic outlet syndrome
	Swelling, pulsation, bruit	
	Adson's sign, hyperabduction test	
Eye	'Horner's syndrome'	Sympathetic involvement – T1 nerve root

Alfred W. Adson (1887–1951). Neurosurgeon, Mayo Clinic, U.S.A. 'Adson's deep breathing test' depends on the fact that the scalenus anterior muscle is an accessory muscle of respiration. A patient with suspected thoracic outlet syndrome is asked to turn his head towards the side of his symptoms. If, when he takes a deep breath and holds it, the radial pulse is diminished or absent, 'Adson's sign' is positive.

Hyperabduction test. Weakness of the muscles normally responsible for holding the clavicle away from the first rib, e.g. sternocleidomastoid, trapezius etc., can cause the subclavian artery to become compressed between the clavicle and the first rib. The test is positive when hyperabduction of the arms above head level cause weakening of the radial pulse.

2. Specific examination of hand tissues

TISSUE	SIGN	SIGNIFICANCE
Skin	The presence or absence of dirt stain, papillary pattern	Use or disuse e.g. nerve or tendon injury
	↓Sweating	Nerve injury
	Tight dystrophic skin ± vascular changes	Collagen diseases, reflex dystrophy
Nail	Distorted growth, grooving, Pitting	Pressure on matrix e.g. scar, mucous cyst Arthritis
	Subungual discoloration ± pain	Splinter haemorrhage Glomus tumour, melanoma
	Clubbing	Chest disease
Blood vessels	Temperature or colour changes, ulcers	Arterial or venous insufficiency
	Diminished wrist and finger pulses	
Fascia	Thickening, nodules, pits	Dupuytren's contracture Palmar fasciitis
Muscle – (intrinsic, – extrinsic)	Wasting, weakness, fasciculation	Nerve palsy Malingering
Tendons	Crepitus, postural abnormality, ↓movement of specific joint, e.g. distal interphalangeal joint/flexor digitorum profundus	Tenosynovitis, tendon division
Nerves – sensory	*Skin.*↓Dirt stain, sensation, sweating, trophic signs, e.g. ulcers	Sensory nerve disorder
– motor	*Muscles.* Weakness, wasting	Motor nerve disorder
Bones. Joints	Deformity, crepitus,↓active or passive movement	Congenital anomaly, injury, arthritis, or neoplasm

3. Examination of overall hand function

Sensation

Motor power (power grip, precision pinch, static hook)

Expression

List of abbreviations

Wherever possible in the book these names have been spelt out in full. Where space did not allow this, the following abbreviations have been used.

Bones
R & U	radius and ulna
C	carpus
MC	metacarpal
PP	proximal phalanx
MP	middle phalanx
DP	distal phalanx

Joints
WJ	wrist joint
MCP	metacarpo-phalangeal
PIP	proximal interphalangeal
DIP	distal interphalangeal

Flexor tendons
FDP	flexor digitorum profundus
FDS	flexor digitorum superficialis
FPL	flexor pollicis longus
FCR	flexor carpi radialis
FCU	flexor carpi ulnaris
PL	palmaris longus

Extensor and abductor tendons
APL	abductor pollicis longus
EPB	extensor pollicis brevis
ECRL	extensor carpi radialis longus
ECRB	extensor carpi radialis brevis
EPL	extensor pollicis longus
EI	extensor indicis
EDC	extensor digitorum communis
EDM	extensor digiti minimi
ECU	extensor carpi ulnaris
LET	lateral extensor tendon (lateral slip)
MET	medial extensor tendon (central or middle slip)
DET	distal extensor tendon
TET	terminal extensor tendon (= DET)
APB	abductor pollicis brevis
FPB	flexor pollicis brevis
OP	opponens pollicis
AP	adductor pollicis

List of tables

1. The normal hand

There is no one example of a normal hand. Each person has individual hand characteristics. Every hand reflects emotional and physical character, and is also a mirror of underlying systemic disease.

A knowledge of basic functional anatomy is essential in diagnosing disorders of the hand. For example, to diagnose the cause of a hand contracture you must know the functional anatomy of the hand, as well as the pathological processes which affect it.

Nomenclature of the hand

	DIGITS	RAYS (Metacarpal and digit)	SURFACES
Thumb	1st	I	Palmar (volar)
Index finger	2nd	II	Dorsal
Middle (long) finger	3rd	III	Lateral (radial)
Ring finger	4th	IV	Medial (ulnar)
Little finger	5th	V	

1. Nomenclature of digits and creases.
- (1) thumb, 1st digit, Ray I
  Ray = digit + metacarpal
- (2) index finger
- (3) middle (long) finger
- (4) ring finger
- (5) little finger, 5th digit, Ray V
- (6) distal interphalangeal joint crease
- (7) proximal interphalangeal joint crease
- (8) palmar digital crease
- (9) distal palmar crease
- (10) proximal palmar crease
- (11) median crease
- (12) thenar crease
- (13) wrist creases

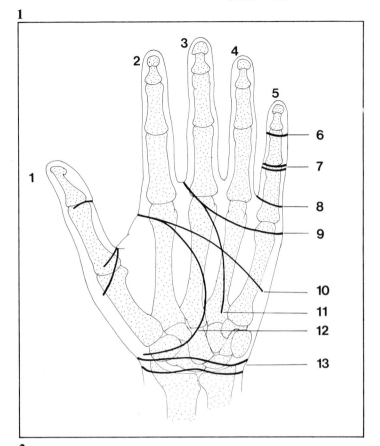

2. Regions and surfaces of the hand.
The palmar (volar) surface – opposite to extensor (dorsal) surface.
- (1) thenar region
- (2) thumb web
- (3) web (interdigital fold)
- (4) palm
- (5) ulnar surface
- (6) hypothenar region

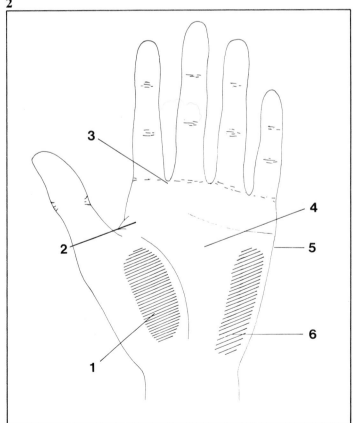

Anatomy and function of individual tissues

Skin

3. Dorsal skin. This is thin (1–2mm), soft, and yielding, and has a loose pliable subcutaneous layer to allow full flexion of the fingers and thumb. Apart from the dorsal aspect of the distal phalanges, this skin contains hair.

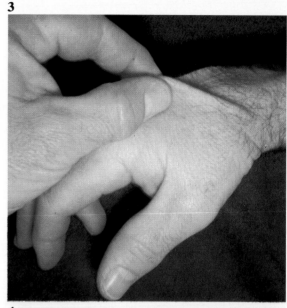

4. Palmar skin. This is thick (4mm) and tough to withstand wear. It covers a thick pad of fat traversed by fibrous septa. It is richly supplied with nerve endings and sweat glands. These factors combined provide a precise pinch and grasp mechanism. A system of creases adhering the skin to the deeper layers allows closure of the hand without the skin bunching up in folds. This skin has no pigment or hair.

Left hand – showing palmar skin and creases.
Right hand – showing pinch (thumb and index finger) and grip (middle, ring and little fingers).

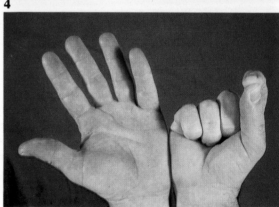

Nails

5. Nails. These are specialised epidermal appendages that support and protect the fingertips and provide a mechanism for picking up and scratching. See **963**.

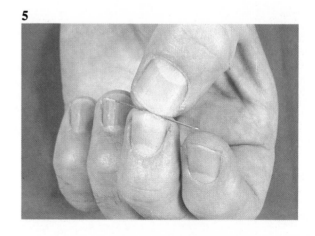

Fascia

6. The palmar aponeurosis is a very strong, somewhat triangular thickening of the central portion of the superficial layer of the deep fascia of the hand. It has longitudinal, transverse and vertical components. It anchors the palmar skin and protects the underlying tendon and neurovascular structures.

(1) palmaris longus
(2) thenar fascia
(3) digital arteries and nerves
(4) fibrous flexor sheaths
(5) superficial transverse metacarpal ligament
(6) transverse fasciculi
(7) palmar aponeurosis
(8) hypothenar fascia
(9) palmaris brevis

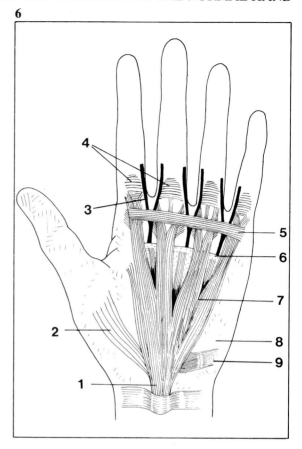

7. Palpating the palmar aponeurosis. Extension of the fingers and thumb makes the fascial bands more prominent (arrows).

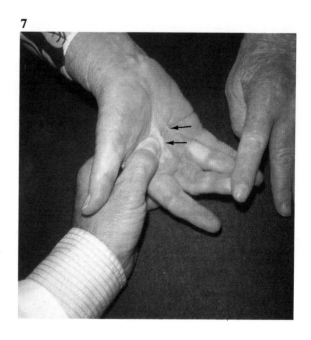

Tendons

FLEXOR TENDONS

8. Palmar view of the flexor tendons. The four flexor digitorum profundi tendons have a common muscle belly in the distal forearm. They insert into the bases of the distal phalanges.

9. The four flexor digitorum superficialis tendons – each has its own muscle belly in the forearm and inserts into the middle phalanges.

10. Flexor digitorum profundus action. Simultaneous flexion of all the distal interphalangeal joints.

11. Flexor digitorum superficialis action. To demonstrate this tendon's independent flexion of the proximal interphalangeal joint, hold the other fingers in extension to block the action of flexor digitorum profundus.

12. Flexor carpi radialis (1) and ulnaris (2) action – demonstrated by forcible extension of the fingers and thumb. The palmaris longus (3) is also shown.

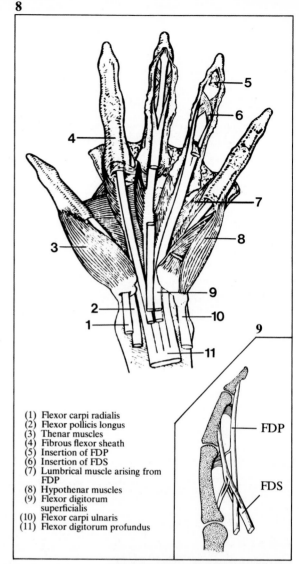

(1) Flexor carpi radialis
(2) Flexor pollicis longus
(3) Thenar muscles
(4) Fibrous flexor sheath
(5) Insertion of FDP
(6) Insertion of FDS
(7) Lumbrical muscle arising from FDP
(8) Hypothenar muscles
(9) Flexor digitorum superficialis
(10) Flexor carpi ulnaris
(11) Flexor digitorum profundus

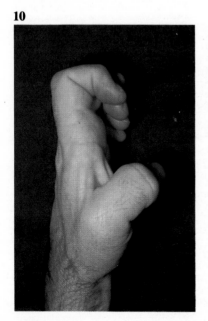

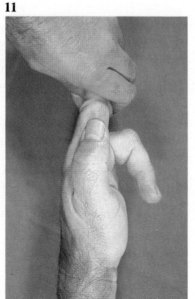

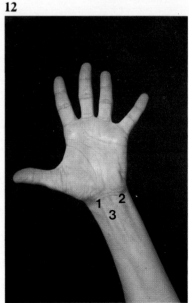

EXTENSOR TENDONS

13. Dorsal view of the hand showing the extensor tendons. Extensor digitorum attaches distally to the extensor hood and combines with the intrinsics to form the digital extensor mechanism. Though it contributes to the extension of all the digital joints, its prime extensor function is at the metacarpo-phalangeal joint.

(1) extensor carpi ulnaris
(2) extensor digiti minimi
(3) extensor indicis
(4) extensor digitorum
(5) intertendinous band (vinculum accessorium)
(6) 1st dorsal interosseous
(7) adductor pollicis
(8) extensor pollicis longus
(9) extensor pollicis brevis
(10) abductor pollicis longus
(11) extensor carpi radialis longus
(12) extensor carpi radialis brevis
(13) extensor retinaculum

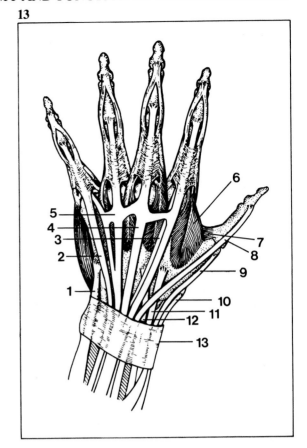

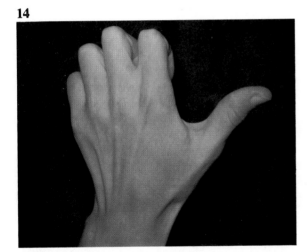

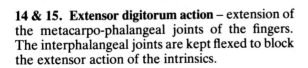

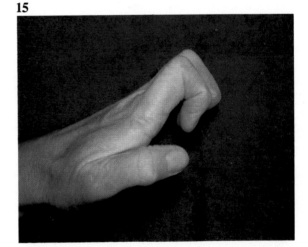

14 & 15. Extensor digitorum action – extension of the metacarpo-phalangeal joints of the fingers. The interphalangeal joints are kept flexed to block the extensor action of the intrinsics.

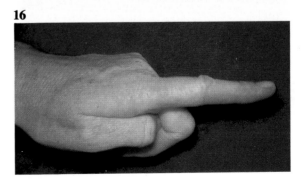

16. Extensor indicis action. Independent extension of the metacarpo-phalangeal joint of the index finger. The other fingers are kept flexed to block the action of extensor digitorum.

In some individuals extensor digitorum can achieve the same function.

17. The radial aspect of the hand and wrist. The 'anatomical snuff box'. Crossing this space are the radial artery and radial nerve.

(1) abductor pollicis longus
(2) trapezium
(3) extensor pollicis brevis
(4) extensor pollicis longus
(5) 2nd metacarpal
(6) scaphoid
(7) extensor carpi radialis longus

Intrinsic muscles

THENAR AND HYPOTHENAR
MUSCLES: opposition and abduction of
the carpometacarpal joint, flexion of the
metacarpo-phalangeal joint and extension of
the interphalangeal joints of the thumb and
little finger.

18. The thenar and hypothenar muscles.

(1) abductor pollicis brevis
(2) opponens pollicis
(3) flexor pollicis brevis
(4) flexor digiti minimi
(5) opponens digiti minimi
(6) abductor digiti minimi

19. Opposition of the thumb to the little finger.
The nails of each digit align in parallel fashion.

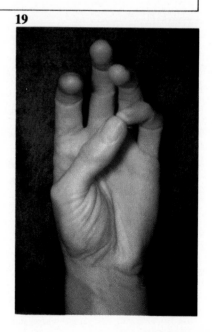

INTEROSSEOUS MUSCLES: abduction and adduction of the metacarpo-phalangeal joints

20

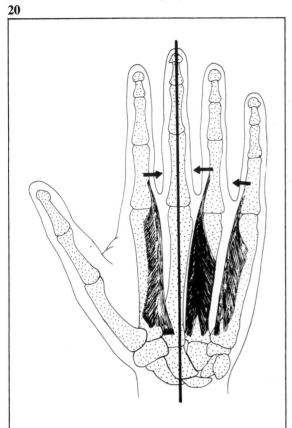

21

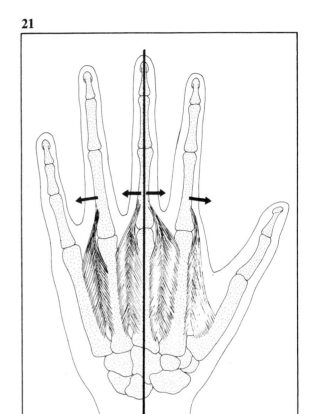

20. Palmar surface of the hand showing the adductor interosseous muscles. ('P.A.D.' The palmar interossei are adductors.)

22a. Interosseous muscle action, testing intrinsic abduction of the middle finger. Both hands are kept flat. Extend the middle finger – this prevents any action by extensor digitorum; abduction of the left middle finger is carried out by intrinsic action only. Ulnar palsy on the right hand prevents intrinsic abduction of that middle finger.

21. Dorsal surface of the hand showing the abductor interosseous muscles. ('D.A.B.' The dorsal interossei are abductors.)

22b. Interosseous muscle action, finger abduction on the right and adduction on the left. The hands must rest on a flat surface. The extrinsic flexors can also function as finger adductors and the extensors as abductors.

22a

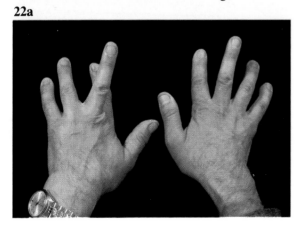

22b

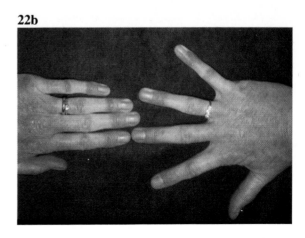

LUMBRICAL MUSCLES: flexion of the metacarpo-phalangeal joints and extension of the interphalangeal joints of the fingers

23 & 24. Diagram of a digit showing action of lumbrical and interosseous muscles, side view and dorsal view.

23 24

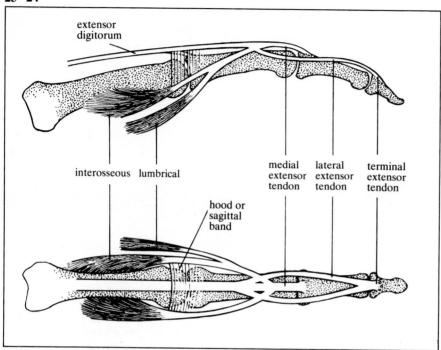

25. Intrinsic muscle action demonstrated. *Left hand* – flexion of the metacarpo-phalangeal joints and extension of the interphalangeal joints. This is sometimes called the 'intrinsic plus' position or deformity.

Right hand – paralysed ulnar innervated intrinsics. This is the 'ulnar claw' deformity. When the median innervated intrinsics are also paralysed the full 'intrinsic minus' deformity is seen.

Mechanism of the 'ulnar claw' hand. Extensor digitorum, unopposed by the paralysed intrinsic flexors of the metacarpo-phalangeal joints, holds these joints in extension. The extrinsic flexors of the fingers, unopposed by the paralysed intrinsic extensors, flex the interphalangeal joints.

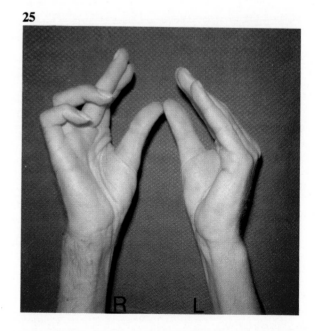

Nerves

26. Sensory nerve supply. The palm on the left, dorsum on the right.
(1) median nerve
(2) ulnar nerve
(3) radial nerve
(x) site to test each nerve

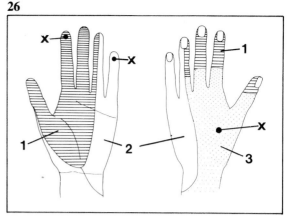

27. Testing pain sensation with a pin. Ask the patient to shut his eyes and keep his hand steady. Ask him to identify the site and nature of the stimulus.

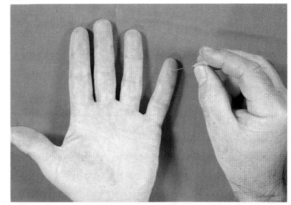

28. Testing touch sensation with cotton wool.

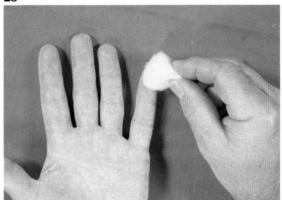

29. The pick-up test for 'tactile gnosis'. The blindfolded patient is asked to pick up and recognise a number of small objects. He does not use the right index and middle fingers, which have no sensation from previous injury.

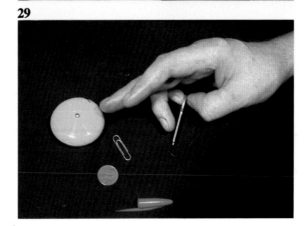

Sweating. Look for sweating on the skin of the pulp through an ophthalmoscope or feel for moisture. Refer to the skin wrinkle sign. See **228**.

30. Absence of sweating on the left from nerve injury. See also **227**.

30

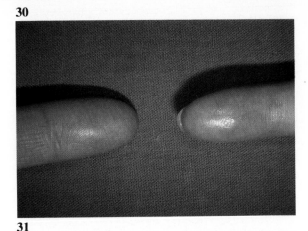

31. Motor nerve supply of the hand*. Two groups of extrinsic muscles: (1) supplied by the radial nerve, (2) supplied by the median nerve, except flexor carpi ulnaris and flexor digitorum profundus (4 + 5) to ring and little fingers.

Three groups of intrinsic muscles: (3) supplied by median nerve, except flexor pollicis brevis†, (4) and (5) supplied by ulnar nerve, except lumbricals (2 + 3).

32. Median nerve supply

33. Ulnar nerve supply. The superficial motor branch leads to abductor digiti minimi, flexor digiti minimi, opponens digiti minimi and palmaris brevis. The deep motor branch to the interossei, the 3rd and 4th lumbricals, adductor pollicis and flexor pollicis brevis.

*These are the most common findings – there are great variations in function of each nerve.

†The superficial head of FPB is supplied by the median and the deep head by the ulnar nerve.

31

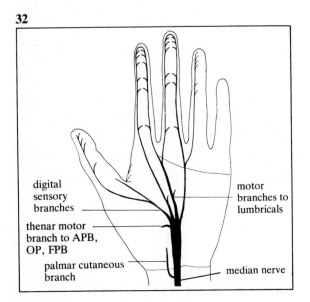

32

33

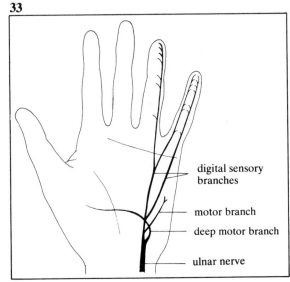

34 & 35. Testing abductor pollicis brevis. This patient had partial division of the left median nerve at the wrist, causing numbness of the thumb and failure of opposition. The instrument points to the paralysed abductor pollicis brevis muscle. See **203, 204**.

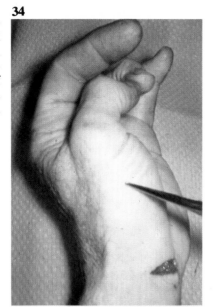

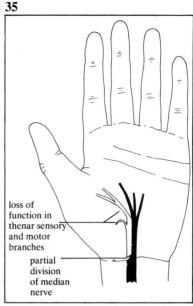

loss of function in thenar sensory and motor branches

partial division of median nerve

36. Testing intrinsic abduction of the fingers against resistance. The right hand shows wasting and claw deformity from ulnar nerve palsy.

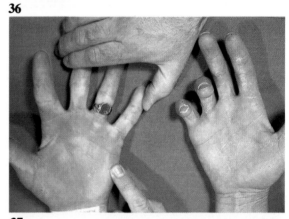

37. Testing intrinsic adduction of the fingers. The 'card test'. Ask the patient to grip a card between his extended fingers. There is weak adduction on the right from ulnar nerve palsy.

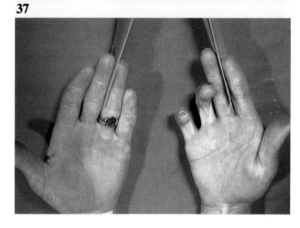

Radial nerve. The radial nerve innervates all the extrinsic, extensor, and abductor muscles of the forearm and hand. See **31**.

38. Wrist drop. This patient had injury to his radial nerve above the elbow. He cannot extend his wrist, fingers or thumb (at the metacarpophalangeal joints). He can extend the interphalangeal joints.

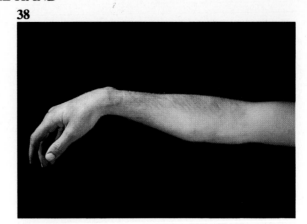

Blood vessels

39. Arteries. The ulnar and radial arteries, by their formation of the superficial and deep palmar arches, and the digital arteries which arise from them, provide the major arterial circulation of the hand.

(1) radial artery
(2) superficial palmar branch
(3) princeps pollicis artery
(4) radialis indicis
(5) proper palmar digital artery
(6) common palmar digital artery
(7) superficial palmar arch
(8) deep palmar arch
(9) ulnar artery

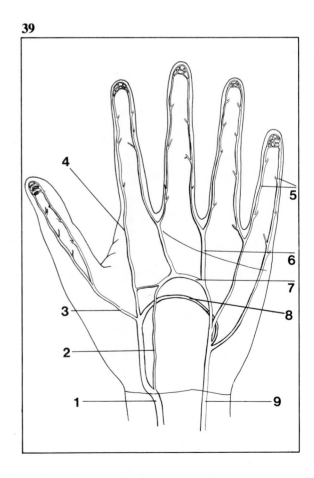

40–43. Allen's test. The patient is asked (**40**) to flex firmly all fingers and the thumb to expel the blood from the hand. Both radial and ulnar arteries are occluded at the wrist. The patient is then asked (**41**) to open his hand. The palm will appear pale. The release of compression of one or other of the arteries at the wrist will result in a rapid return of normal colour to the palm and fingers, provided that the vessel is patent. If the vessel is not patent, pallor will persist.

40

41

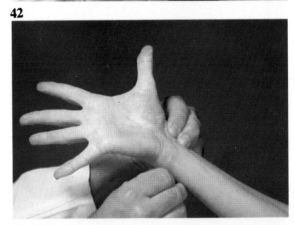

42. Release of ulnar artery compression. The hand remains pale.

42

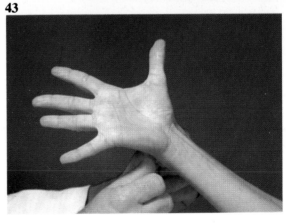

43. Release of radial artery compression. The hand regained its pink colour. The radial artery is supplying most of the blood to the hand.

43

44. Veins. The major veins and lymphatics are predominantly on the dorsum of the digits of the hand where they cannot be compressed with gripping. There is, however, a deep network of fine veins on the volar aspect, and with each artery are companion venae comitantes.

(1) basilic vein
(2) dorsal venous arch
(3) superficial dorsal veins
(4) intercapitular veins
(5) dorsal digital veins
(6) perforating vein
(7) cephalic vein

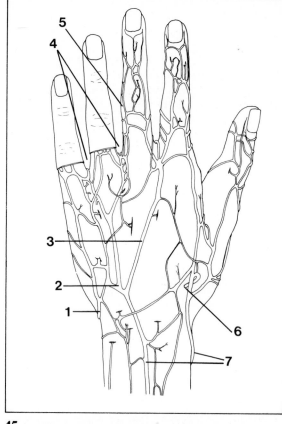

45 & 46. Testing circulation of the nail bed and pulp. Pressure on the nail bed or pulp with a probe or pencil will cause immediate blanching, while release of pressure is followed by an almost instantaneous return of the pink colour. This is not a reliable test of the circulation as it can also be demonstrated after death.

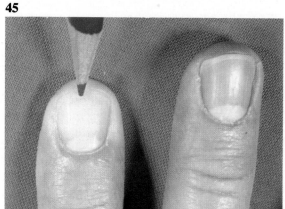

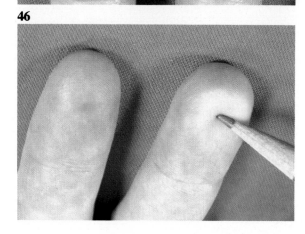

Lymphatics

There is a superficial system which follows the venous drainage and a deep system which accompanies the arteries.

47. The lymphatics on the dorsum of the hand.

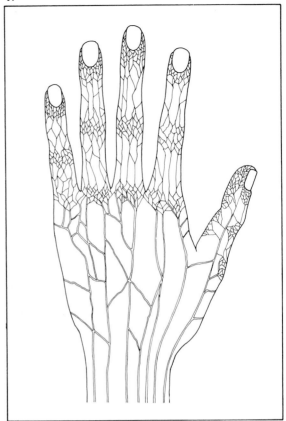

Bones, joints and ligaments

48

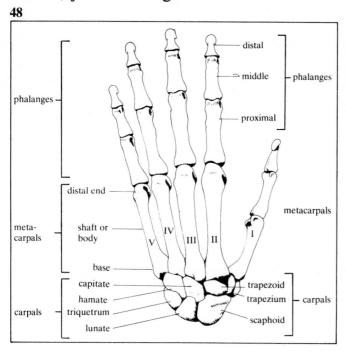

49

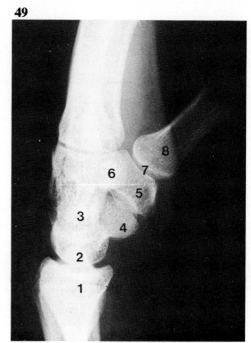

48. Bones of the hand, dorsal view.

49. Radiograph of the left wrist from the lateral side, with the thumb abducted. (1) lower end of radius, (2) lunate, (3) capitate, (4) pisiform, (5) trapezium, (6) hook of hamate, (7) carpo-metacarpal joint of thumb, (8) first metacarpal.

50

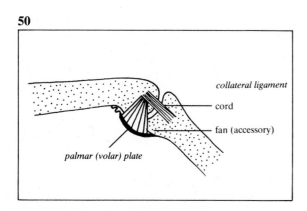

51

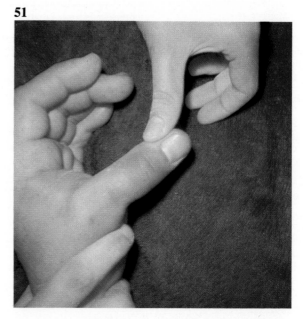

50. The metacarpo and interphalangeal joint ligaments. Each joint has lateral (collateral and fan) and palmar ligaments (palmar plate). The 'fan' ligament is the accessory collateral ligament.

51. Testing stability of a metacarpo-phalangeal joint of the thumb. The joint should be stable in extension. Here there is rupture of the ulnar collateral ligament with instability of that joint and free passive abduction.

Testing overall hand function

The arm has one function: to place the hand where it is needed. The hand has three functions*.

(1) Sensory perception ('eyes') and sensory contact (soothing, social and sexual).

(2) Expression (direction, demonstration and dancing).

(3) Mechanical grips (precision manipulation, static hook and power grip).

Functionally, sensation is more important than movement.

Handedness. The right hand is dominant in 80–90% of individuals, but neither hand is dispensable. Left handedness is associated with speech and reading difficulties. It is also more common in genius and criminals.

Sensation

52. Testing sensation. To undo a button hidden from view requires sensory discrimination as well as dexterity. See also **217**.

Movement

PINCH

Precision manipulation. Dependent on the intrinsics for both posturing and movement, but the extrinsics provide stability and strength. The median nerve is important as it provides discriminatory sensation and innervation of the thenar intrinsics – these keep the thumb anterior to the plane of the hand.

53. Precision pinch. The tips of the fingernails of the thumb and index finger are opposed.

54. Key pinch. Pulp of thumb against radial side of middle phalanx of index finger.

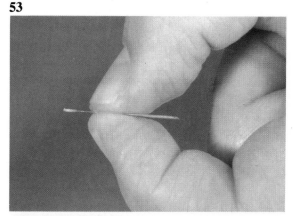

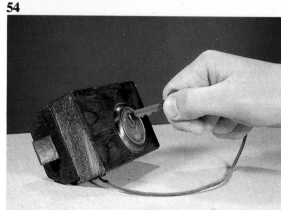

* *J. T. Hueston.* Contemporary Australian plastic and hand surgeon.

GRIP AND GRASP

Grip and grasp mainly depend on the extrinsic muscles. The extensors posture the hand preparing for grip and the flexors clamp the 'vice' of the fingers against the palm. The strength of grip is proportional to the range of flexion in the smallest joints. The sensory requirements are only protective and proprioceptive.

55. Hook grip – carrying a briefcase. The fingers are extended at the metacarpo-phalangeal joints and flexed at the interphalangeal joints.

56. Power grip – using a hammer. The fingers are flexed and the opposed thumb can be flexed over the fingers to gain extra power.

57. Grasp and bilateral hand function. Such tasks as lifting heavy weights or clapping require two hands. The thumbs are fully abducted and the fingers are locked in semi-flexion.

58. Chuck grip. The importance of the dominant hand. The pulps of the index and middle fingers and thumb are opposed to hold a pen like the chuck in a drill.

55

56

57

58

EXTENSION

59. The flat hand – as in pushing against a flat surface or placing the hand between closely approximated surfaces. It is also used in crawling, e.g. by miners. The fingers and thumb are extended at all joints. The thumb is adducted to lie in the same plane as the hand.

EXPRESSION

60. The prayer position. The hand is an organ of expression. Both hands are needed for this and other forms of expression such as clapping.

Surface and applied anatomy

61. Palmar surface markings. An oblique line is drawn from a point 1cm distal to the pisiform bone across the palm to the thumb web.

A vertical line is extended along the radial border of the middle finger to intersect the oblique line. This point represents the recurrent motor branch of the median nerve. A vertical line is extended along the ulnar border of the ring finger. It intersects the oblique line at the hook of the hamate. At this point the ulnar nerve divides into the deep motor and superficial sensory branch.

The deep palmar arch lies just proximal to the oblique line. The superficial palmar arch is at the level of the thumb web.

(1) flexor carpi radialis
(2) radial artery
(3) deep palmar arch
(4) median nerve and its thenar branch
(5) superficial palmar arch
(6) ulnar nerve and its deep branch
(7) pisiform
(8) ulnar artery

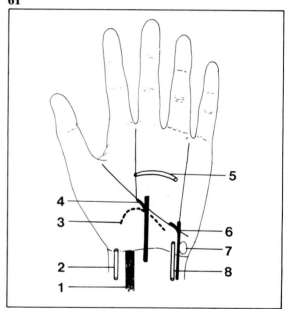

62. Dorsal landmarks

(1) head of ulna
(2) extensor pollicis longus
(3) styloid process at the distal end of the radius
(4) Lister's tubercle

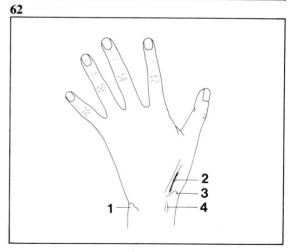

Variations in the normal hand

Age and sex

Growth of the Hand. At birth, the hand is fully developed except for the nervous system. Myelinization is not complete until the age of two. Thus, such subtle neurological tests as stereognosis, see Chapter 3, are of no value in infants.

Hand length nearly doubles in the first two years of life, and nearly doubles again in the remainder of the growth period.

Babies' fingers are conical; the tips are rounded and the fingernails are very small. The flexibility of a baby's fingers is due to supple ligaments. Ossification starts between the first and second years. The carpal wrist bones do not begin to form until the second year and do not fully develop until the ninth year. The fatty tissue that covers the fingers and hand of a baby causes dimpling at all knuckles.

Until puberty, a girl's and boy's hand look the same. From that time on the girl's hand becomes more delicate and the boy's hand tougher.

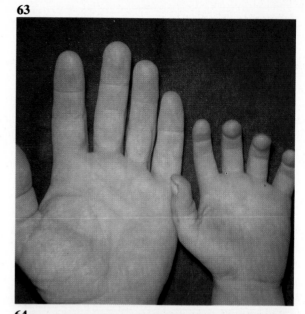

63. Comparison of an adult's and a five-year-old child's hand. See also **134**.

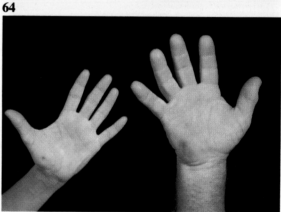

64. Two different sizes of adult hand: the right hand of a farmer and the left hand of his wife.

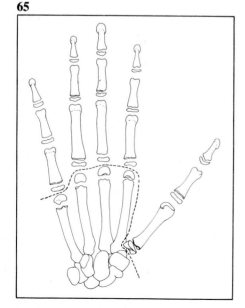

65. The epiphyses of the hand. The phalanges have a single proximal epiphysis. The 1st metacarpal behaves as a phalanx, having a proximal epiphysis. The other metacarpals have distal epiphyses.

66. Carpal ossification. Beginning at the capitate which ossifies at about one year the carpal bones ossify in the sequence shown. The approximate age in years is indicated on each bone. Until the age of eight the number of ossification centres corresponds with the child's age.

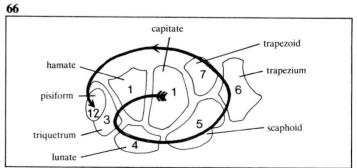

67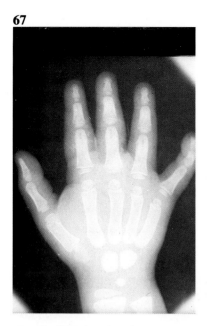

68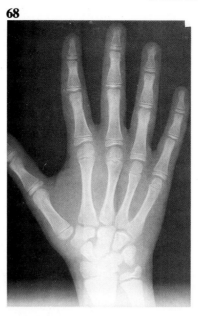

69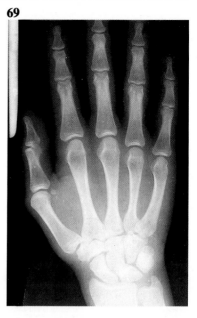

67–69. Hand and wrist x-rays. A four-year-old child (67). The capitate, hamate, triquetrum and lunate are ossified. The distal radial epiphysis is present.

68. A twelve-year-old. All carpal bones are ossified but the epiphyses have not fused. The centre of ossification of the pisiform usually appears at this stage.

69. An eighteen-year-old. Growth almost complete. Finger and hand epiphyses have fused. The arrow points to the fused radial epiphysis.

The carpal bones have assumed adult shape.

Occupation

70. Rough and smooth hands. The smooth clean skin of a clerical office worker compared to the rough dirt-stained skin of a labourer's hand. The labourer is wearing a copper bangle in the hope of preventing arthritis.

70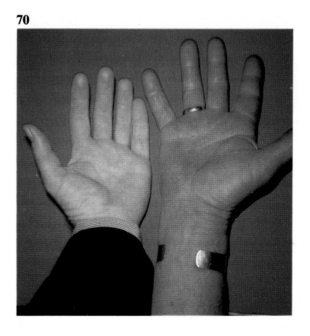

Build

71. A thin hypermobile hand with ligament laxity. This is common in oriental and negroid races. See also **876**.

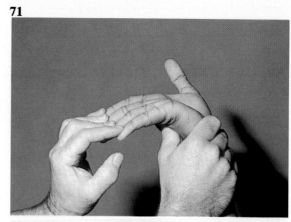

72. A thick hand. Thick palmar fascia prevents passive hyperextensibility. See also **7**.

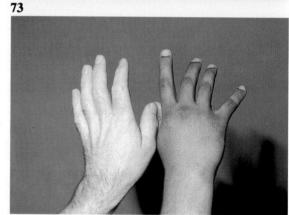

Race

73. White and coloured hands. The white Caucasian compared to the darker skin of an Indian.

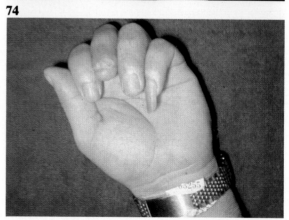

Religion and culture

74. Long fingernail. The little fingernail may be kept long for cosmetic, religious or practical reasons, e.g. cleaning the ear (digit auricularis).

2. Congenital disorders

A short glossary of words and prefixes used to describe congenital disorders

cheir- Gk hand, also spelt *chir-*
dactyl-, dactylos Gk finger, toe, digit
mel-, melos Gk limb, member

a- without
aplasia atrophy or absence (*a-*, Gk *plassein* to form)
acro- extremity, peak (Gk)
arachno- spider (Gk)
brachy- short (Gk)
campto- bent in flexed position (Gk *kamptos* bent)
clino- bent in lateral deviation (Gk *klinein* to bend)
ectro- abortion, absence (Gk *ektrosis* miscarriage)
hemi- half (Gk)
megalo, macro- great, large, long (Gk)
micro- small (Gk)
oligo- few, deficient (Gk)
pero- deformed (Gk)
phoco- seal (Gk)
poly- many, much (Gk)
syn-, sym- together (Gk)

See also the Appendix, page 332, for a fuller glossary of terms used in congenital disorders.

Congenital disorders of the hand and upper limb can be classified into six main groups. See **75**.

They can present as a limb bud deficiency or a systemic disorder e.g. heart/hand syndrome. Two glossaries of terms used to describe congenital disorders are shown on pages 35 and 332.

Incidence. About 0.8% of live births are associated with congenital disorders of the upper limb. The commonest disorders are polydactyly and syndactyly.

Aetiology. There are three groups of causes.

(1) Intrinsic or endogenous i.e. genetic (dominant, recessive or sex linked). This may be hereditary. See below.

(2) Extrinsic or exogenous (i.e. environmental) e.g. nutrition, trauma, chemical (thalidomide), infection (rubella), irradiation, toxic, endocrine.

(3) A combination of intrinsic and extrinsic factors.

The type of abnormality depends not so much on the specific agent, but on the stage of development. Most abnormalities are produced during the period of rapid differentiation, which is in the first seven weeks of embryonic development.

Significance. If the first born child is defective, the birth of a subsequent malformed child is 25 times more likely. Some conditions are hereditary e.g. syndactyly, polydactyly, clinodactyly, symphalangism. Some are associated with systemic disorders e.g. heart/hand syndrome (congenital heart disease – radial club hand), head/hand syndromes e.g. Apert's syndrome. Some are associated with foot disorders e.g. syndactyly, cleft hand. Age of parent: the older the mother, the higher the incidence of mongolism.

Eugene Apert (1868–1940). French paediatrician. Described Apert's syndrome, i.e. acrocephalosyndactyly – a combination of face and hand deformity, in 1906.

(1) Failure of formation of parts (arrest of development, absence defects)

A. Transverse

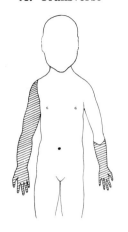

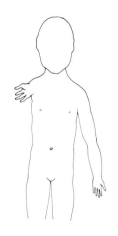

B. Longitudinal

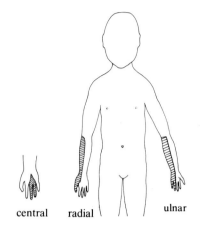

central   radial   ulnar

**(a) transverse terminal
(amelia)**

**(b) transverse intercalary
(intermediate)**

**(a) longitudinal terminal
(b) longitudinal intercalary**

**(2) Failure of differentiation
of parts**

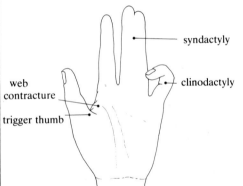

syndactyly

clinodactyly

web
contracture

trigger thumb

(3) Focal defects

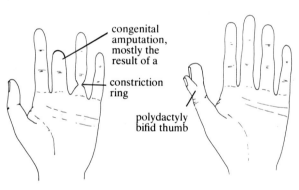

congenital
amputation,
mostly the
result of a

constriction
ring

polydactyly
bifid thumb

(4) Duplication

(5) Overgrowth (gigantism)

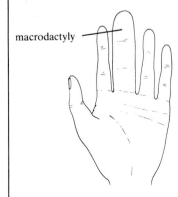

macrodactyly

**(6) Generalised skeletal
defects affecting
the upper limb**

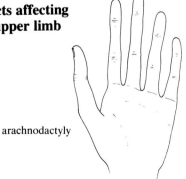

arachnodactyly

1. Failure of formation of parts (arrest of development, absence defects)

These absence defects are usually exogenous, due to some accident in inter-uterine devélopment e.g. thrombosis, leading to tissue death, amputation, and absorption.

A. Transverse

(a) Transverse terminal
There may be absence of the terminal phalanges or complete absence of the entire limb (amelia). In these 'congenital amputations' the stump is usually well padded. The rudimentary digits may show as dimples.

76. X-ray of amelia. Absent right limb and rudimentary left upper limb.

77. Acheiria. Congenital amputation through the forearm.

78. Acheiria and adactyly. Congenital amputation through the carpus on the one hand and congenital amputation through the bases of the index, middle, and ring fingers on the other hand.

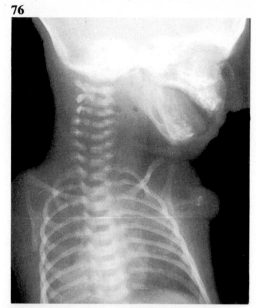

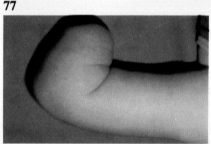

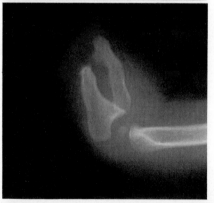

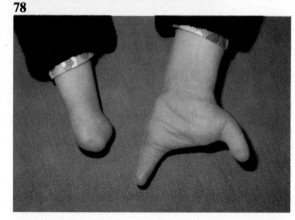

79

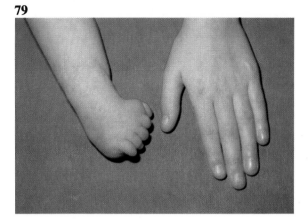

79. Rudimentary fingers. Amputations at the carpo-metacarpal region.

80

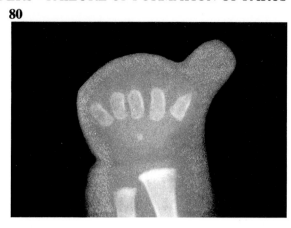

80. Adactylia. Absence of fingers.

81

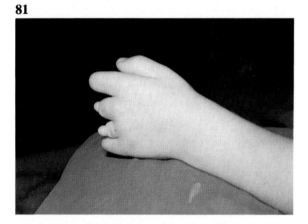

81 & 82. Aphalangia. Absent middle and distal phalanges and short proximal phalanges.

82

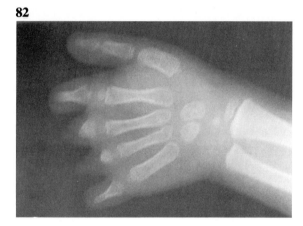

83

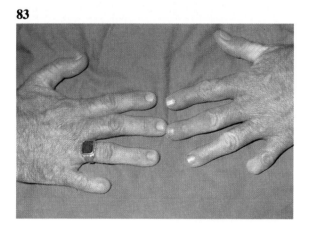

83. Aphalangia. Absent distal phalanges in each little finger of a 50-year-old man.

84

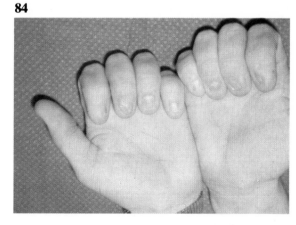

84. Anonychia. Congenital absence of the finger-nails. Nail beds present.

For other nail deformities Chapter II.

(b) Transverse intercalary (intermediate)

This type is sometimes referred to as 'phocomelia' because of its resemblance to a seal's flipper. There may be arrest of the upper arm and forearm (hand attached to trunk), the upper arm (hand and forearm attached to trunk), or the forearm (hand attached to elbow).

This type of deformity is commonly seen in thalidomide babies.

85 & 86. Phocomelia. Hand attached to trunk.

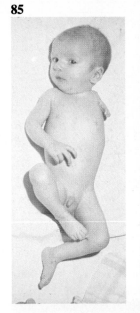
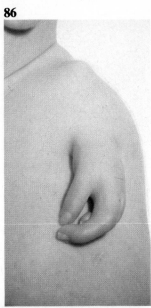

87. Phocomelia. Hand attached to elbow. X-ray of a similar case.

88. Congenital short forearm. Absent elbow, gross ectromelia and ectrodactyly.

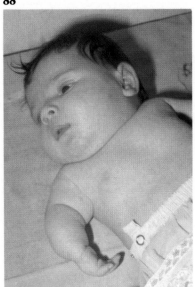

B. Longitudinal

There are three developmental areas of the fore-arm and hand.

(i) The radial component forming the radius, thumb and intervening carpal bones.

(ii) The central component comprising the index and middle fingers and central carpal bones.

(iii) The ulnar component comprising the ulna, ulnar two fingers and their carpal bones.

89. The three developmental areas of the embryonic upper limb – (1) Radial, (2) Central, and (3) Ulnar.

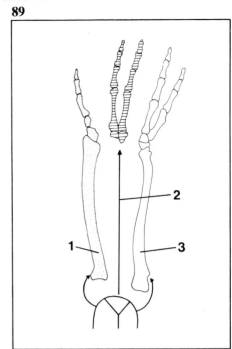

(a) Longitudinal terminal

90. Radial. Absent thumb.

91 & 92. Central. Lobster, cleft or split hand.

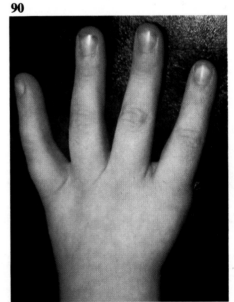

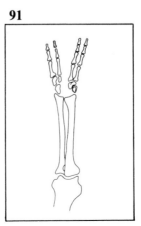

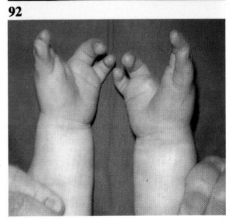

93–95. Ulnar arrest of development. The left hand shows ulnar terminal loss (little and ring finger), and the right hand shows a severe reduction deformity.

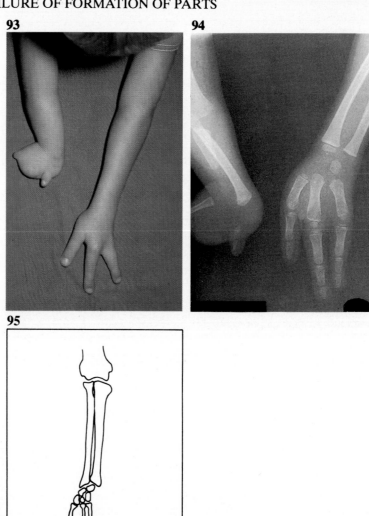

(b) Longitudinal intercalary
Longitudinal loss of an intermediate part which can be radial, or ulnar (this is rare).

96. Radial.

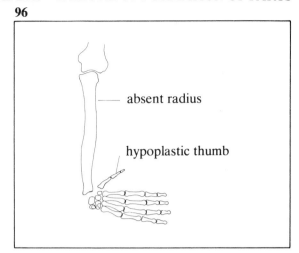

97 & 98. Absence of the radius, i.e. 'radial club hand' or 'hemimelus' in a Spanish child. Radial deviation of the forearm and wrist and hypoplasia of the thumb ray.

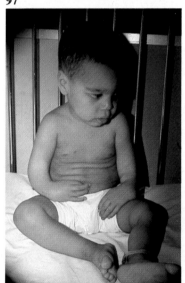

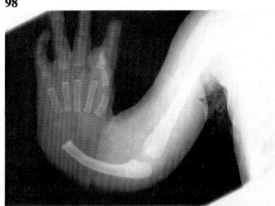

99. 'Pouce flottant'. Because of hypoplasia of the first metacarpal, the thumb literally floats on the radial side of the hand.

100. X-ray of 'pouce flottant'. In this case there were two phalanges, a nail and good sensation.

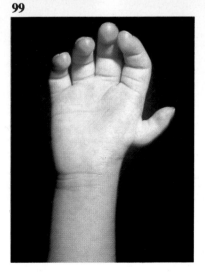

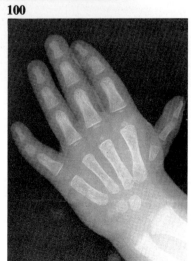

2. Failure of differentiation of parts

Shoulder, arm and forearm

101 & 102. Congenital elevation of the shoulder.
Sprengel's deformity. One shoulder is elevated
and the highly placed scapula is usually tilted towards the spine. Often associated with skeletal
anomalies of the cervical and thoracic spine; can
be associated with absence of the pectoralis major
muscle, as shown here, or breast.

101

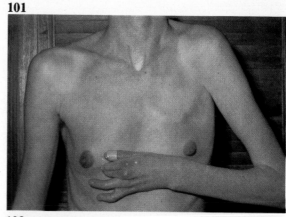

102

103. X-ray of Sprengel's deformity on the left side.

103

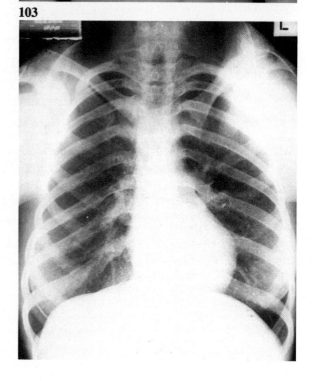

Otto Sprengel (1852–1915). German general surgeon. Described in 1891
the deformity which is called by his name, i.e. failure of the scapula to
descend from its original high position in embryonic life.

104

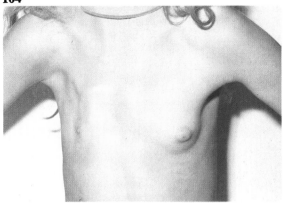

105

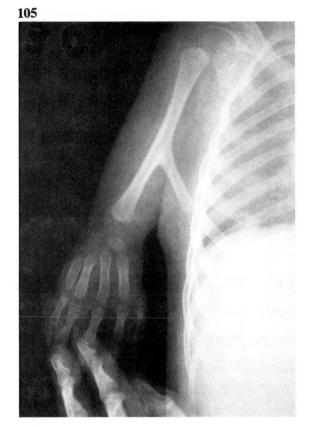

104. Poland's syndrome. Absence of the pectoral muscles associated with hypoplasia of the nipple and aplasia of the breast. For associated syndactyly see page 47.

105. Synostosis of the elbow. Usually associated with arrest of development of the ulnar component.

Hand

HAND. Carpal deformities rarely cause problems by themselves when the rest of the hand is normal.

106. Bipartite scaphoid – division of the scaphoid may be partial or complete. It may be the result of two ossification centres or the result of injury.

106

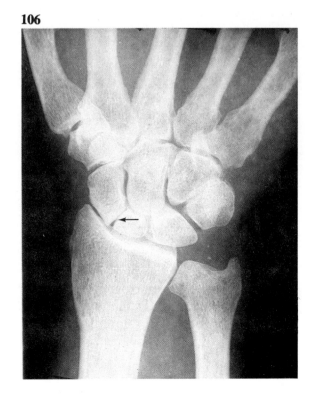

Alfred Poland (1820–1872). English surgeon. Described brachy-syndactyly associated with the absence of the pectoral muscle which he noted in a cadaver whilst he was demonstrating in anatomy at Guy's Hospital.

107

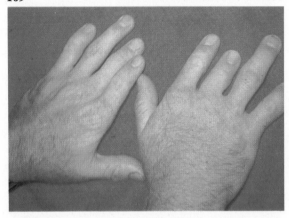

108

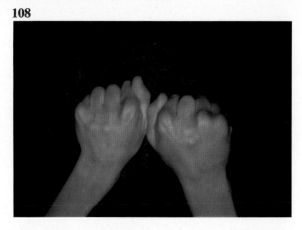

Brachydactyly. Short fingers from short phalanges or metacarpals. The first example of mendelian inheritance demonstrated in man. Often associated with syndactyly or other digital deformity.

107 & 108. Brachydactyly. Short right middle and left middle and ring fingers from arrest of metacarpal development. A 12-year-old boy. Normal hand function.

109

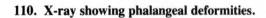

110

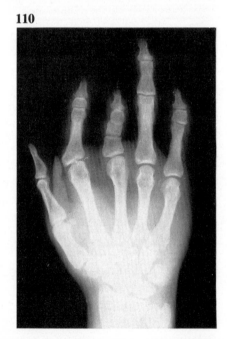

109. Short right middle and left index and middle fingers from arrest of phalangeal development. Normal function.

110. X-ray showing phalangeal deformities.

111. Stub thumb. Short distal phalanx of thumb.

111

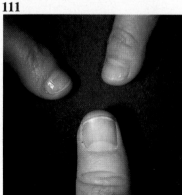

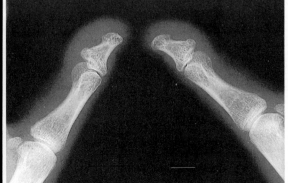

Digit

SYNDACTYLY

Syndactyly is fusion or webbing of the fingers; it can be congenital or acquired.

Congenital syndactyly is the second most common hand malformation, occurring once in every 2–3,000 births. It is twice as frequent in males. It usually involves the middle/ring finger cleft but can involve any of the others, though rarely the thumb/index finger cleft. It is often symmetrically bilateral, can affect the toes and is often associated with other deformities of the same upper limb, including hypoplasia and polydactyly. It is also associated with other congenital malformations, e.g. Apert's syndrome (when associated with acrocephaly), and Poland's syndrome (when associated with absence of the pectoralis major muscle).

It is familial, often dominant and repeats itself from one generation to the next. A family history of a similar deformity can be obtained in 20–30% of cases.

Classification. Syndactyly can be classified as complete or incomplete, simple or complicated.

Complete syndactyly extends to the tip of the digit. *Incomplete syndactyly* falls short of that. *Simple syndactyly* involves only the skin. *Complicated syndactyly* means tissues other than the skin are fused, e.g. bone, tendons, nerves, and nails. *Acrosyndactyly* is when the ends of the fingers are joined together.

Eugene Apert (1868–1940). French paediatrician. Described Apert's syndrome, i.e. acrocephalosyndactyly – a combination of face and hand deformity, in 1906.

Alfred Poland (1820–1872). English surgeon. Described brachysyndactyly associated with the absence of the pectoral muscle which he noted in a cadaver whilst he was demonstrating in anatomy at Guy's Hospital.

112. Simple incomplete syndactyly between the right middle and ring fingers. A corresponding syndactyly on the left hand was corrected by surgery.

113. Simple complete syndactyly extending along the length of the middle and ring fingers. This patient, under anaesthetic, had his hand held flat for the purpose of this photograph.

112

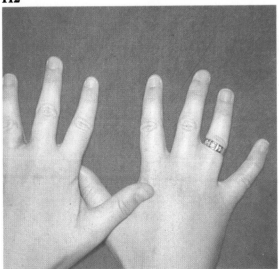

113

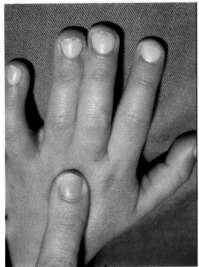

114. Complicated syndactyly with fusion of the bony parts ('mitten hand').

115. Mitten hand, complicated syndactyly.

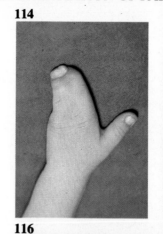

116. X-ray of a similar case.

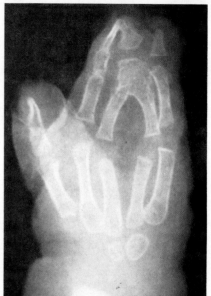

117–119. Complicated syndactyly of the hands and feet, associated with craniosynostosis.

117

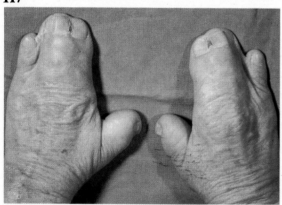

118

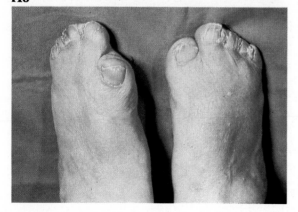

119

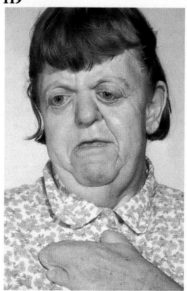

120. Mitten hand with thumb syndactyly.

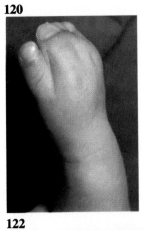
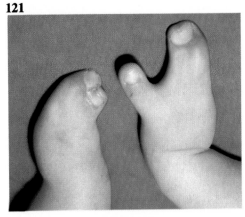

121. Oligosyndactyly. Syndactyly associated with loss of digits.

122. Syndactyly associated with polydactyly.

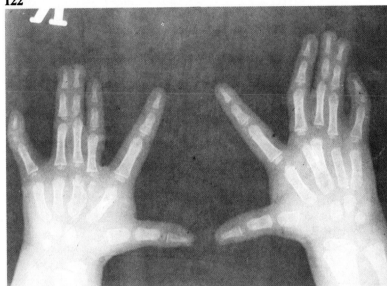

123. Syndactyly associated with brachydactyly.

124. Acrosyndactyly. The ends of the fingers are fused together.

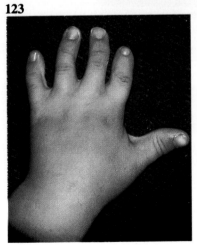
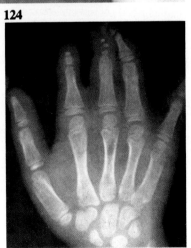

SOFT TISSUE CONTRACTURES

Contractures of the soft tissues such as the skin, fascia, ligaments or muscles of the hand can present as congenital trigger thumb or finger, camptodactyly, congenital contracture of the thumb web or arthrogryposis.

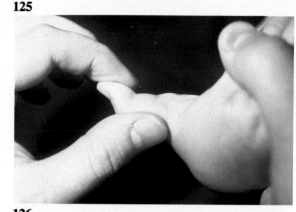

125. Congenital trigger thumb in a three-month-old infant. Notta's node; a nodule on flexor pollicis longus catches on the fibrous flexor sheath and prevents extension of the metacarpo-phalangeal and interphalangeal joints of the thumb.

126. Camptodactyly. A painless and usually progressive bilateral flexion contracture of the proximal interphalangeal joints, usually of the little finger. It can be mild, without interference of function, or severe, with increasing secondary flexion contracture of all structures on the volar side of the joint. Hereditary. See also **625**.

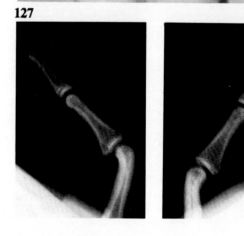

127. X-ray showing flexion contracture of the proximal interphalangeal joint, tapering of the head of the proximal phalanx and subluxation of the base of the middle phalanx.

128

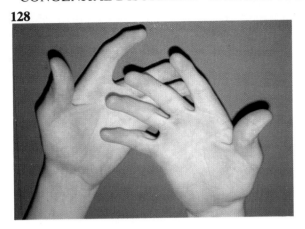

129

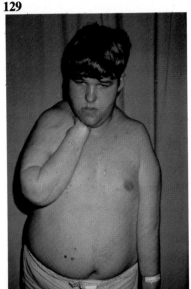

128. Congenital contracture of each thumb web.

129. Arthrogryposis. In this condition there is absence of individual muscles or groups of muscles, producing stiffness of the joints they control. There is an increase in the subcutaneous fat and webbing of the neck and axilla.

SKELETAL DEFORMITIES

Here there is a failure of development of part of a bone resulting in unequal growth or fusion of neighbouring bone.

130. Madelung's deformity. Deficiency of the ulnar half of the distal radial epiphysis leads to a relatively excessive length of the ulna, and secondary derangement of the distal radio-ulnar joint.

130

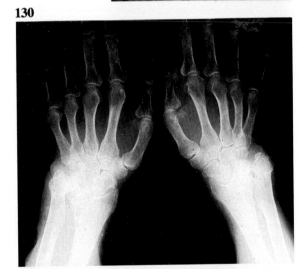

131. Madelung's deformity. This 65-year-old female did not present until the secondary arthritis at her wrist caused rupture of the extensor tendons to her ring and little fingers.

131

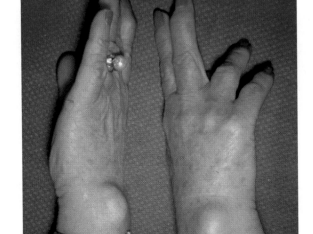

Otto Wilhelm Madelung (1846–1926). German professor of surgery. In 1878 he described the deformity of the wrist occurring in young women, probably due to a defect in the development of the distal radial epiphysis.

132. Clinodactyly. Bent fingers in lateral deviation. The little finger has lateral and flexion deviation at the proximal inter-phalangeal joint. Clinodactyly is usually bilateral and symmetrical, and may be due to unequal growth of the two sides of the epiphysis, to malalignment of the joint, or to delta phalanx. See also **626**.

133. X-ray showing lateral and flexion deviation at the proximal interphalangeal joint. Often there is hypoplasia and incurving of the middle phalanx.

132

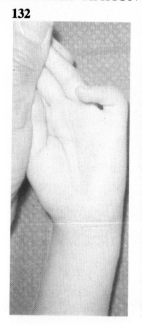

133

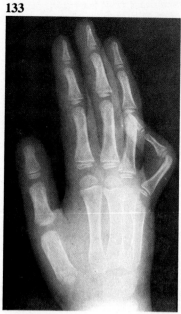

134. Clinodactyly of the left index finger in a six-month-old child.

134

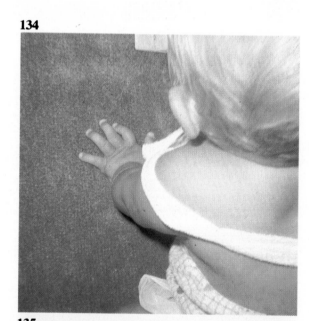

135. Symphalangism (stiff fingers). There is a developmental stiffness, usually in extension, because of bony or fibrous union. This patient cannot flex the proximal interphalangeal joint of his little finger. A dominant type of mendelian inheritance; often associated with syndactyly.

135

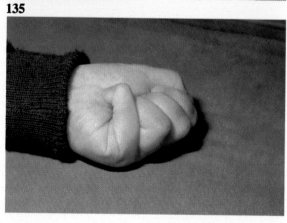

3. Focal defects

Necrosis of foetal mesenchyme may develop into annular constricting bands or rings. These may cause intra-uterine gangrene and true foetal amputation. The remaining parts may fuse, forming acrosyndactyly.

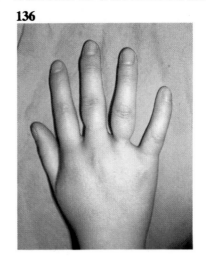

136. Annular constricting bands in a 25-year-old Korean woman. These bands were superficial

4. Duplication

Digital duplication, e.g. polydactyly. The most common of all hand anomalies. It is frequently hereditary and associated with other deformities such as syndactyly and brachydactyly. The supernumerary digits are usually marginal, occurring on the radial (pre-axial) or ulnar (post-axial) side of the hand. The condition is often bilateral, making six digits on each hand. Duplication can be partial or complete, and can occur at the level of the metacarpal or the phalanx. The extra digit can have independent action.

137. Bifid thumb, at the distal phalanx level.

138. Bifid thumb. One metacarpal and two separate phalangeal components.

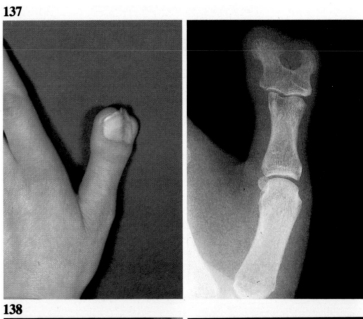

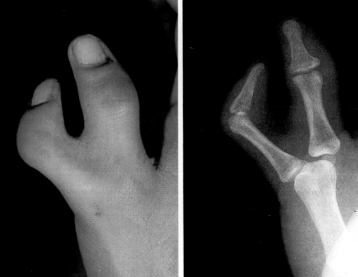

139

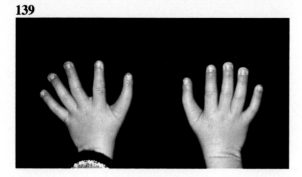

140

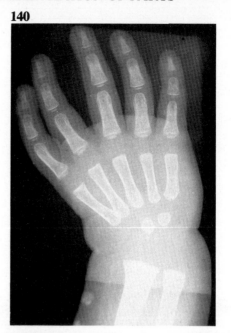

139 & 140. A five-fingered hand. The thumb is triphalangeal and non-opposable. The web space between the two radial digits is narrow.

Mirror hand. Mirror hands are nearly identical, each being the mirror complement of the other. Even the carpus may be double. The thumb and thenar eminence are usually absent.

141. Mirror hand. Duplication of the ulna and the ulnar side digits. The right hand is normal.

141

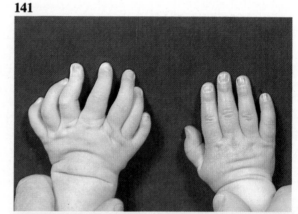

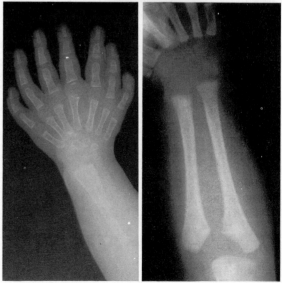

5. Overgrowth (gigantism)

Gigantism may affect all or part of the limb and it may involve the skeletal as well as the soft tissue elements.

Macro or megalo dactylia. This usually involves the index finger and thumb, is seldom hereditary and is not associated with other deformities. It is usually a hyperplastic overgrowth of both skeletal and soft tissue components of the finger, but not the metacarpal.

The soft tissue hyperplasia consists of fat, neurovascular and lymphatic elements. There may be an associated neurofibroma, lymphangioma or haemangioma.

142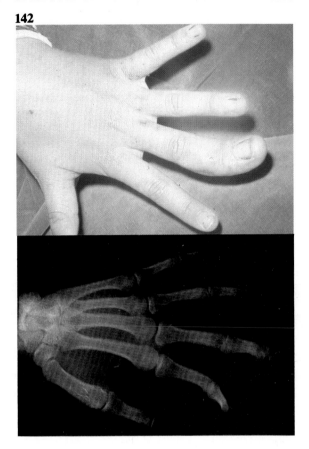

142. Macrodactyly in the left middle finger of a 10-year-old Italian girl. The x-ray shows enlarged skeletal and soft tissue structures.

143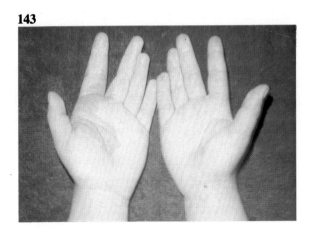

143. Gigantism in the left index finger of a 6-year-old Greek child. There is an associated haemangioma.

144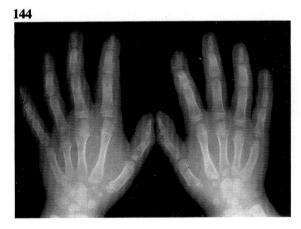

144. Enlarged metacarpal and phalanges of the left index finger. X-ray.

6. Generalised skeletal defects affecting the upper limb

Here there are congenital malformations of the upper limb associated with systemic skeletal defects such as dyschondroplasia, diaphyseal aclasis, arachnodactyly and achondroplasia.

145 & 146. Dyschondroplasia, Ollier's disease, multiple chondromatosis. Ossification of cartilage at the growth discs is faulty, with islands of cartilage remaining unossified within the shaft and not on its surface as in diaphyseal aclasis. The fingers show multiple enchondromata.

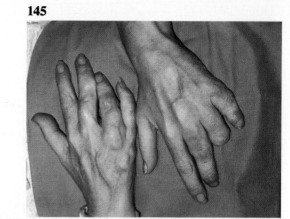

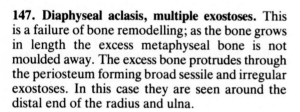

147. Diaphyseal aclasis, multiple exostoses. This is a failure of bone remodelling; as the bone grows in length the excess metaphyseal bone is not moulded away. The excess bone protrudes through the periosteum forming broad sessile and irregular exostoses. In this case they are seen around the distal end of the radius and ulna.

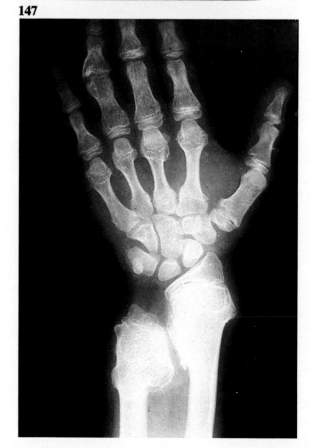

Louis Ollier (1830–1900). French orthopaedic surgeon. Described Ollier's disease, i.e. dyschondroplasia in 1899.

148

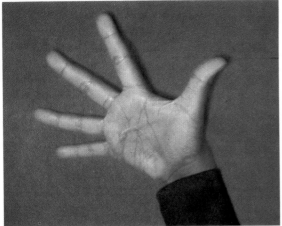

149

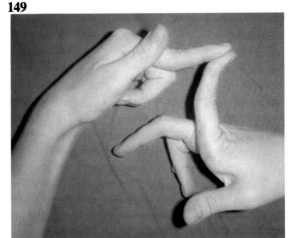

148 & 149. Arachnodactyly. Spider fingers in an Indian adult female. The metacarpals and phalanges are long, giving the fingers a spidery look. There is often associated ligament laxity and systemic disorders which include congenital heart disease, infantilism and scoliosis. There may be an underlying endocrine disorder.

150a, b. Achondroplasia. This is a disorder of cartilage growth. If the child survives he becomes a dwarf with grossly short limbs. The fingers are short and thick.

151. Fragilitas ossium, osteogenesis imperfecta, or brittle bones in a 15-year-old girl. The fractured radius has been fixed by a wire. She had previously had multiple other fractures. There was bone rarefaction on x-ray.

150a

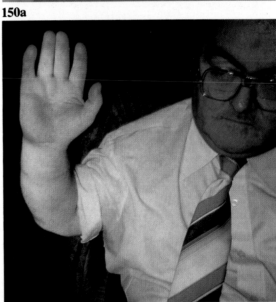

150b

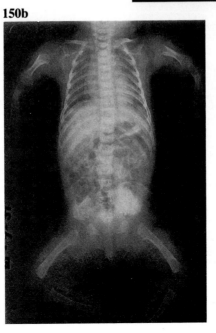

151

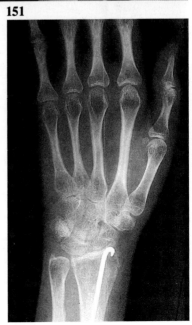

3. Injuries

General classification

Open injuries
tidy wounds 152–154
untidy wounds 155

Closed injuries 156–158

Complex hand injuries
avulsion and mangle 159–165
high pressure injection 166–170
gun shot and blast 171–174

Tendon injuries (open)

Flexor
complete 175, 176
partial 177

Extensor
terminal extensor tendon 178
medial extensor tendon (central slip) 179
extensor digitorum 180–183
wrist and digit extensors 184, 185

Intrinsic
dorsal interosseous 186
palmar interosseous and lumbrical 187

Tendon injuries (closed)

Flexor
in the digit 188–190
at the wrist 191, 192

Extensor
terminal extensor tendon 193, 194
extensor pollicis longus 195, 196
extensor digitorum, 'Vaughan Jackson'
    syndrome 197, 198
extensor hood 199, 200

Nerve injuries

anatomy of a peripheral nerve 201
microscopic types of nerve injury 202

Open; median nerve, partial, complete 203–206

Closed; median, ulnar, median and ulnar 207–213

The clinical signs of peripheral nerve injury
pathology – nerve degeneration and
    regeneration 214
sensory signs 215–220
motor signs 221–226
autonomic signs 227–230
miscellaneous signs 231–233

Specific patterns of peripheral nerve injury

injuries to the brachial plexus
upper level, Erb's palsy 234, 235
upper and mid level 236–238
mid and lower level, Klumpke's palsy 239, 240
total brachial plexus palsy 241–243

radial nerve injury
at the elbow 244, 245
in the upper forearm 246

median nerve injury
at the elbow 247–255
at the wrist 256

ulnar nerve injury
at the wrist 257–265

combined median, radial and ulnar nerve injury
    266–268

digital nerve injury 269

Vascular injuries

Acute
crush injury 270–277

Chronic
hand hammer syndrome 278, 279
digital hammer syndrome 280, 281
Raynaud's phenomenon 282

Traumatic aneurysm and arteriovenous fistula
    283

Iatrogenic and self-inflicted injuries 284–289

Miscellaneous
ischaemic contracture 290, 291
spontaneous haemorrhage 292

Skeletal injuries

Joint injuries
sprain of proximal interphalangeal joint 293, 294
dislocation of distal interphalangeal joint 295, 296
joint injury with ligament rupture 297–300
minute intra-articular fracture at base of middle
    phalanx 301
intra-articular fracture of proximal
    interphalangeal joint with secondary joint
    dislocation 302
semi-lunar, and peri-lunar dislocation,
    complicated by median nerve compression 303

Fractures
tendon forces acting on a fractured proximal
    phalanx 304
transverse fracture with angulation deformity
    305, 306
spiral fracture of proximal phalanx with
    rotational deformity 307, 308

Fingertip injuries

Open
levels of open fingertip injury 309
partial skin loss 310
complete skin loss 311, 312
partial avulsion of pulp 313
complete avulsion of pulp 314

Closed
crush injury with distal pulp haematoma 315, 316
extensive subungual haematoma 317
dislocation of nail apparatus 318, 319
mallet finger 320, 321

Foreign bodies

tumour from glass foreign body 322–324
cystic tumour from fish scale 325, 326
foreign body reaction to wooden splinter 327
foreign body reaction to catgut suture 328
steel pin foreign body 329, 330
x-ray diagnosis 331, 332
sinus due to glass foreign body 333, 334
nerve division from glass fragment 335
ramset nail foreign body 336
fish hook foreign body 337
pyogenic granuloma, see 463

Burns

types of burn 338

Superficial
flame 339
electric flash 340
steam 341
hot fat 342–344
hot tar 345
lime 346

Deep
lysol 347
hot rollers 348
high voltage electricity 349, 350
friction 351
radiation 352
the aftermath of burns 353

Miscellaneous injuries

Self-inflicted
suicide laceration 354, 355
tight band 356–358
puncture or inoculation 359
chemicals 360
drug injection 361
tattoos 362, 363
hysterical palsy 364

Bites
snake bites 365, 366
frost 'bite' 367

General classification

Hand injuries and wounds can be classified as open, closed, and complex.

Open injuries

These may be tidy or untidy.

TIDY WOUNDS

These are usually caused by a knife or broken glass, and result in a clean cut incised wound with minimal contamination and a good blood supply.

152. Tidy wound (arrow) of the left middle finger from grasping a knife. The posture of the finger suggests divided flexor tendons. See **154**.

153. Failure of flexion of the distal interphalangeal joint and painful weak flexion of the proximal interphalangeal joint. This posture also suggests flexor tendon injury.

154. The associated deep damage, which includes a completely divided flexor profundus tendon and a partially divided flexor superficialis tendon and digital nerve. The digital artery was also divided.

(1) flexor digitorum superficialis
(2) digital artery
(3) digital nerve
(4) flexor digitorum profundus

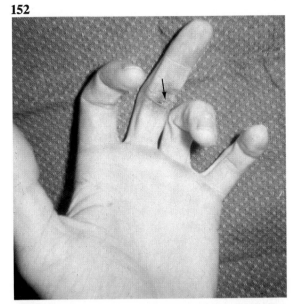

152

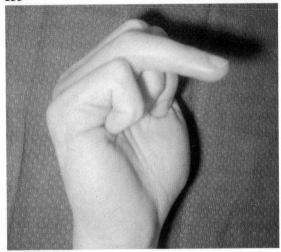

153

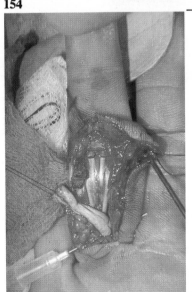

154

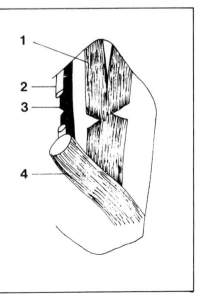

UNTIDY WOUNDS

These are ragged, torn, and caused mostly by machinery. The wound edges are poorly perfused and often contaminated.

155. Untidy wound of a hand caught in a printing press.

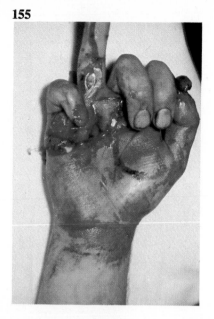

Closed injuries

156. Closed crushing injury of a fingertip which was jammed in a closing door. Note the skin contusion.

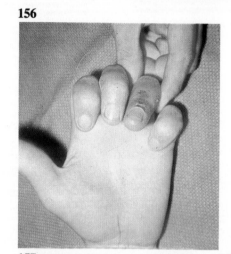

157 & 158. Associated fracture of the distal phalanx and associated closed rupture of the flexor profundus tendon. The tissue forceps indicate the scar, which fills the space previously occupied by the flexor tendon.

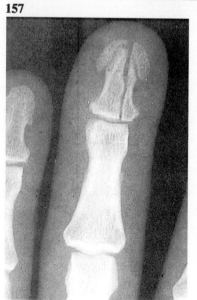

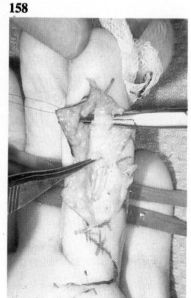

Complex injuries

These include amputations, avulsions, mangles, burns, blast, and injection injuries, all of which involve extensive deep damage to multiple tissues and potential loss of circulation.

AVULSION AND MANGLE

159. Complex hand injury from mangling in a printing machine. Ink stains cover the hand.

160. Mangling from an auger machine.

161a, b. Degloving injury. This process worker had his right dominant hand caught in a roller-type machine. There is stripping of the skin and soft tissues from his fingers. Palmar view, and dorsal view at operation.

162. Avulsion amputation of all fingers which were caught in a machine. One flexor tendon has been avulsed from its muscle tendon junction in the forearm.

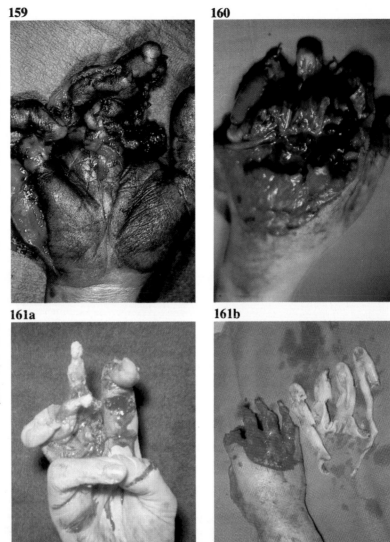

159 **160**

161a **161b**

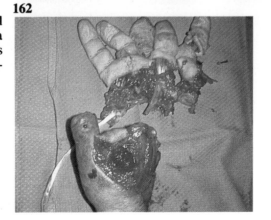

162

163 & 164. Ring avulsion injury. The force of injury is equivalent to the body weight being transferred to that area around the base of the finger. The looser and sharper the ring the more severe the injury. Picture **163** shows incomplete circumferential soft tissue damage – distal segment of finger viable. Picture **164** shows complete circumferential avulsion of soft tissues with profound ischaemia of the distal segment.

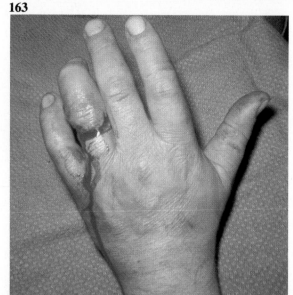

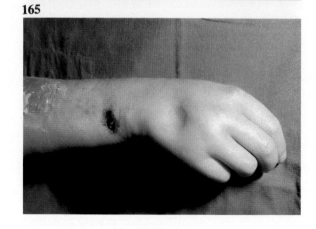

165. Wringer or mangle injury with severe swelling. This patient went on to develop Vokmann's ischaemic contracture. See **614**.

HIGH PRESSURE INJECTION INJURY

When gases, liquids, paint or grease are injected into the hand under high pressure, the combined effect of high pressure and toxaemia in closed tissue spaces produces cellulitis, vasculitis, thrombosis, and reactive fibrosis.

In the first few hours after injury from a high pressure source, the only clinical feature may be a painless puncture wound in the pulp. The deep damage is always more extensive than anticipated.

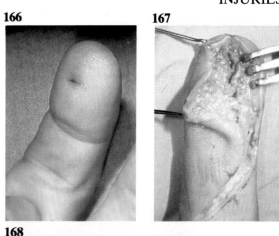

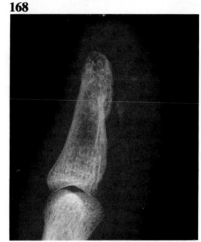

166. Paint gun injury. This patient had an accidental high pressure injection of paint when he used his left index finger pulp to clean the nozzle of a paint gun. This picture shows the pulp five hours after injury. There was, by this time, increasing throbbing pain.

167. Operative finding six hours after injury – necrosis of pulp tissue.

168. X-ray – paint in the pulp tissues.

169. Ischaemic necrosis and infection found at operation 48 hours after high pressure injection of grease.

170. Gangrene – four weeks after high pressure injection of oil.

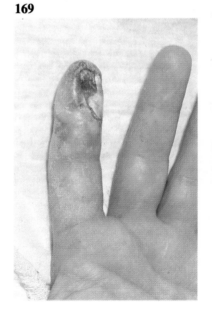

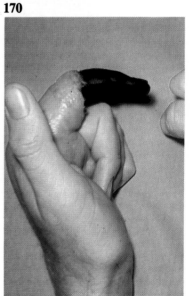

GUN SHOT AND BLAST INJURIES

The degree of injury depends on the velocity, type, and range of missile and degree of contamination from clothing.

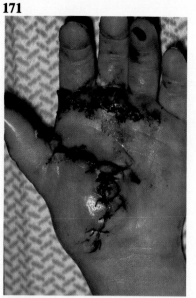 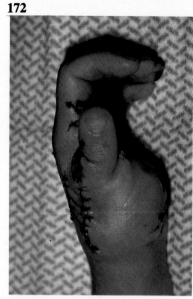

171 & 172. Blast injury from explosion of a 0.22mm cartridge held in the hand. Suturing of the wound aggravated the tendency to severe swelling.

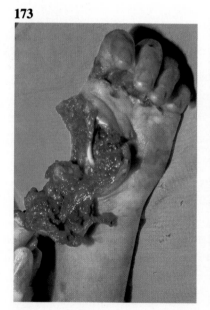 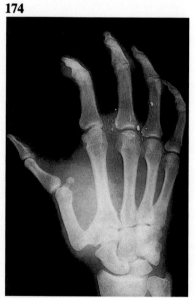

173. The degree of deep damage found at operation one day later. Severe laceration of all tissues on the volar side of the thumb – skin, muscles, tendons, nerves, blood vessels, ligament, and bone.

174. X-ray. Dislocation of the carpo-metacarpal and metacarpo-phalangeal joint of the thumb, soft tissue swelling of the thenar web and cartridge fragments in the hand.

Tendon injuries (open)

Tendon injuries may be open or closed. Open tendon injuries may be complete or partial.

Flexor

175. Complete division of flexor digitorum profundus and superficialis after the patient gripped broken glass. Failure of proximal and distal interphalangeal joint flexion.

176. The divided flexor digitorum profundus at operation.

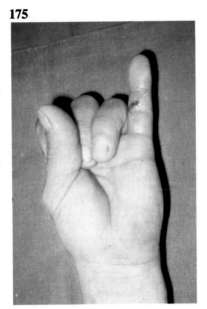

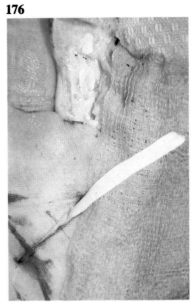

177. Partial division of flexor digitorum profundus (arrow) and complete division of flexor digitorum superficialis. Partial flexor tendon injury presents as painful weak flexion. The diagnosis can be made only at operation. See **154**.

EXTENSOR

178. Division of the terminal extensor tendon of the left middle finger causing flexion deformity of the distal interphalangeal joint. The patient could passively but not actively extend this joint. 'Mallet' or 'baseball' finger.

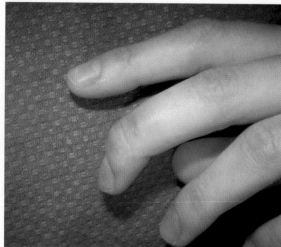

179. Division of the medial extensor tendon, i.e. the central slip. Flexion deformity of the proximal interphalangeal joint. This may take 2–3 weeks to develop. 'Boutonnière' deformity.

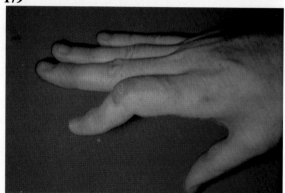

180. Division of extensor digitorum to the right middle finger from a cut on the dorsum of the hand. Note sagging of the finger. The intact intrinsics maintain extension of the interphalangeal joints.

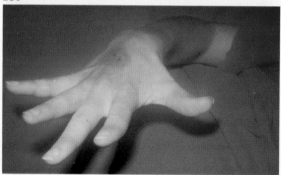

181. Division of extensor digitorum from a knife wound. The index and middle fingers are being extended by extensor indicis and extensor digitorum minimi, which were not divided.

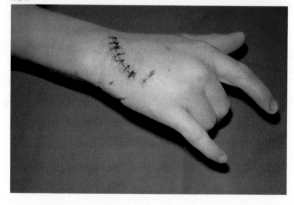

182. Open wound caused by a circular saw over the metacarpo-phalangeal joints.

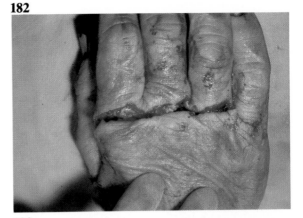

183. Operative findings. Division of the skin, extensor tendon, and capsule of the metacarpo-phalangeal joint.

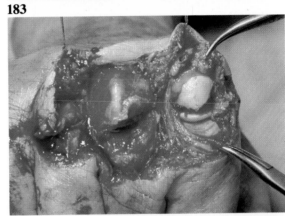

184. Open wound over the extensor aspect of the right wrist.

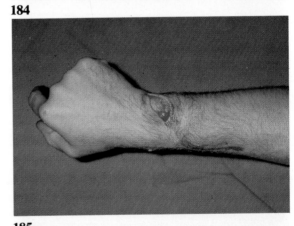

185. Divided extensor tendons. The glass fragment penetrated obliquely, dividing all extensor tendons to the wrist and digits. This patient can make a fist but he cannot extend his wrist, fingers or thumb.

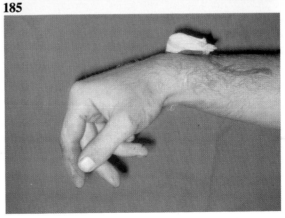

Intrinsic

186. Division of the first dorsal interosseous tendon over the index metacarpo-phalangeal joint. The imbalance of tendon forces causes the finger to lie in ulnar deviation.

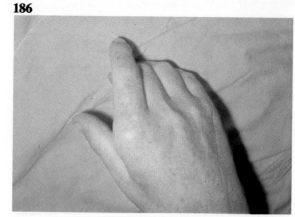

187. Division of lumbrical and palmar interosseous tendons in the palm with failure of full intrinsic extension of the fingers. The arrow points to the wound.

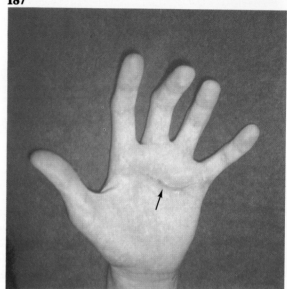

Tendon injuries (closed)

Flexor

188. Closed rupture of flexor profundus of the right ring finger in a rugby player who grasped the jersey of a runaway opponent. He was unable to flex the distal interphalangeal joint and had weak flexion of the proximal interphalangeal joint. The avulsed profundus tendon retracted, forming a painful swelling at the base of the finger, and blocking flexor digitorum superficialis function.

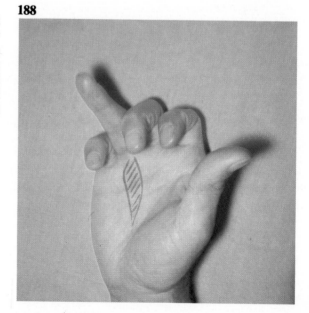

189. Lack of full extension from scar entrapment of flexor digitorum superficialis.

190. X-ray showing avulsed bony insertion of flexor digitorum profundus which has retracted to the proximal interphalangeal joint.

Closed rupture of flexor profundus from direct crushing injury (see **158**).

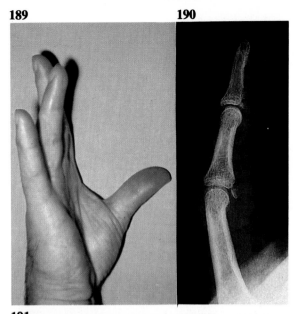

191. Partial closed rupture of flexor digitorum profundus of the left little finger. This woman, a keen golfer, hit hard ground with her club, fracturing the hamate bone. Rubbing of the tendon against the fracture caused it to rupture some weeks later.

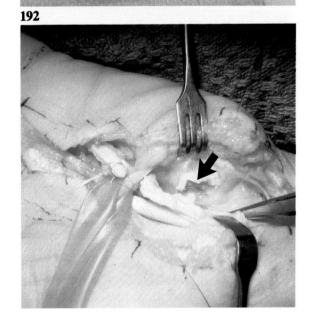

192. Operative finding. Irregular bony surface at the site of fracture with attrition of the flexor profundus (arrow).

Extensor

193. Closed rupture of the terminal extensor tendon of the little finger following forced flexion injury of the distal interphalangeal joint with the fingers extended, as can occur in making a bed or catching a ball. 'Baseball' or 'mallet' finger.

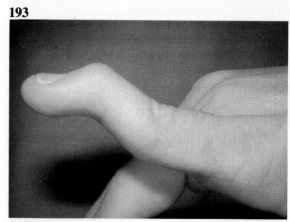

194. A small fragment of bone avulsed from the base of the distal phalanx. X-ray.

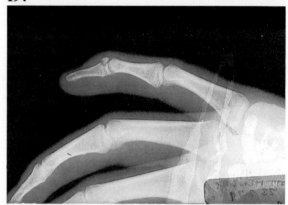

195 & 196. Closed rupture of extensor pollicis longus of the right thumb. This can occur spontaneously when there is roughening of the bony canal beside Lister's radial tubercle, for example, in rheumatoid arthritis of the wrist or after a Colles' fracture. Note the inability to hyperextend the interphalangeal joint of the right thumb and inability to raise the thumb to the plane of the hand. The finger joints show osteo-arthritis.

Though not apparent in this picture, extensor pollicis brevis should extend the metacarpo-phalangeal joint of the thumb.

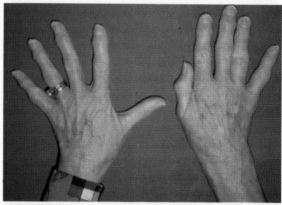

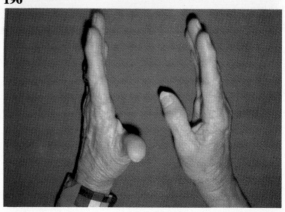

Abraham Colles (1773–1843). Irish surgeon. Professor of anatomy and surgery. Described the common type of fracture of the lower end of the radius in 1814.

197

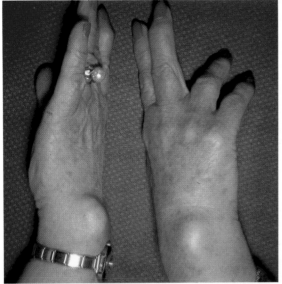

198

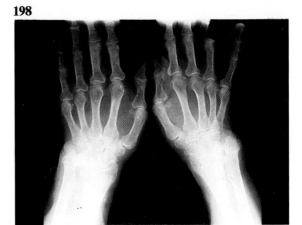

198. Abnormal projection of the distal end of each ulna.

197. 'Vaughan Jackson' syndrome, (i.e. spontaneous closed rupture of extensor tendons to the right ring and little fingers from bone irregularity over the distal end of the ulna). The most common cause is rheumatoid arthritis but this patient had Madelung's deformity.

Oliver James Vaughan-Jackson. Contemporary English surgeon.

199 & 200. Closed rupture of the extensor hood (sagittal band) of the left middle finger. This 60-year-old male attempted to grasp a fly-away bird. Swelling of the soft tissues around the metacarpo-phalangeal joint. He could flex this joint but the ruptured radial part of the hood allowed the extensor digitorum to fall from its central position onto the ulnar side of the joint, thus preventing full extension.

(1) rupture of 'hood'
(2) extensor digitorum
(3) dislocation of extensor digitorum

199

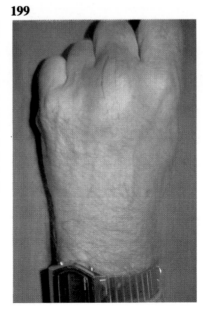

200

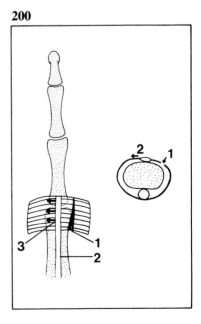

Nerve injuries

201. Anatomy of a peripheral nerve.

Types of nerve injury. These may be classified as:

(a) Microscopic – neuropraxia, axonotmesis, neurotmesis.
(b) Macroscopic – closed or open, partial or complete.

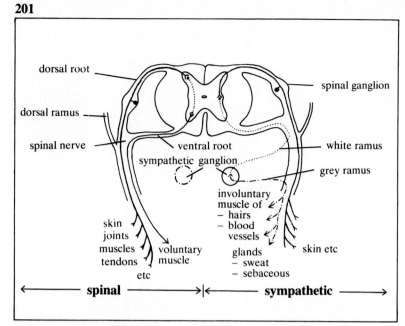

202. Microscopic types of nerve injury. (1) *neuropraxia* – concussion. (2) *neurotmesis* – sheath intact, axons severed. (3) *axonotmesis* – sheath and axons severed.

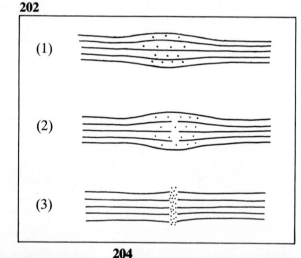

Open

Always suspect a nerve injury in any wound of the forearm or hand. If a wound causes arterial bleeding in the finger there is almost certainly associated nerve damage because the artery lies deep to the nerve.

Nerve damage may be partial or complete.

203. Testing abductor pollicis brevis. This patient had partial division of the left median nerve at the wrist. He had numbness of the thumb and index finger and failure of opposition. The instrument points to the paralysed abductor pollicis brevis muscle. See **35**.

204. Operative findings – partial division of the median nerve (arrow).

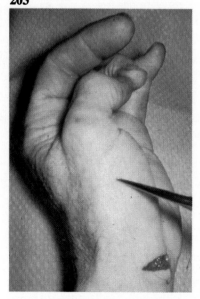

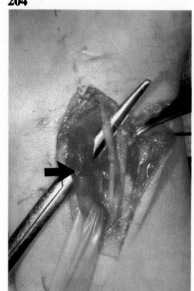

205. Complete division of the median nerve from open injury five years before. Note the swelling of the proximal neuroma, the wasting of the abductor pollicis brevis and the area of numbness (median nerve distribution).

206. Operative findings – proximal neuroma with scar bridging the gap between the two ends of the nerve.

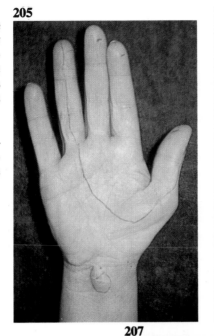

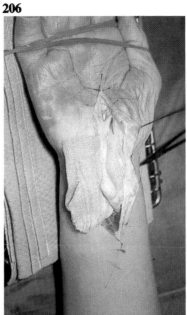

Closed

207. Closed injury of the median nerve. Soon after a heavy steel plate fell on this man's wrist, he noticed numbness of his left thumb and index finger and weakness of abduction, opposition and pinch.

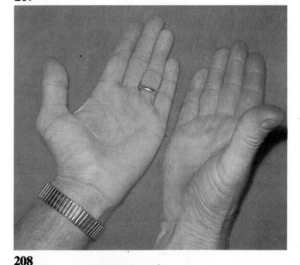

208. Operative findings – bruising (neuropraxia) of the median nerve.

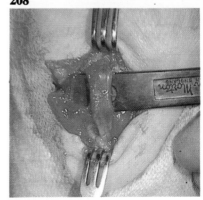

209

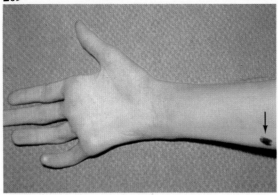

210

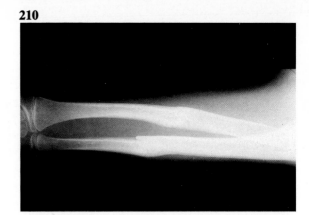

209. Right ulnar nerve palsy complicating a closed fracture of the ulna. Intrinsic muscle weakness and wasting and loss of sensation in the distribution of the ulnar nerve. The arrow marks the site of the fracture and of the Tinel sign over the ulnar nerve. At operation the nerve was found to be completely divided, i.e. neurotmesis.

210. Fracture of the right ulna (after reduction).

Jules Tinel (1879–1952). French physician. Described a test for regenerating sensory axons. The test does not become positive until six to eight weeks after a wound or repair of a peripheral nerve. Then, percussion over that site produces a tingling pain.

211

213

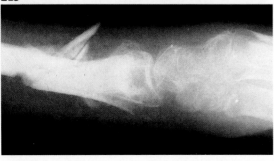

212

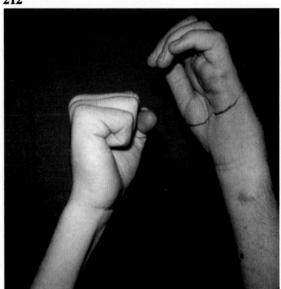

211–213. Median and ulnar nerve palsy complicating a closed comminuted fracture of the lower radius and ulna. Attempted extension (**211**), attemped flexion (**212**) and x-ray (**213**).

This 18-year-old postman fell off his bicycle.

Clinical features of peripheral nerve injury*

NERVE FUNCTION	SYMPTOM	SIGN
Sensory		
Touch	Numb, painful, burning	Diminished dirt staining of skin Loss of touch sensation
Protection (pain, heat, cold)	,, ,, ,,	Blister, loss of pin-prick sensation
Stereognosis	,, ,, ,,	Difficulty in picking up and recognising objects e.g. shape, texture etc Also – Hyperaesthesia, Tinel's sign
Motor		
Early	Weak	Weakness
Late	Very weak and stiff	Wasting, deformity, and contracture
Autonomic		
Sweat	Numb and dry	Dry (absent sweating), absent skin wrinkling
Sebum, hair, nail pigment	,, ,, ,,	,, ,, ,,
Vasomotor		
i Vaso-dilatation	,, ,, ,,	Pink and warm
ii. Vaso-constriction	,, ,, ,,	Blue and cold

*In cases of ACUTE nerve injury the only signs may be loss of pin-prick sensation, muscle weakness, and lack of sweating. The only symptom may be altered sensation.

Key to the diagnosis of a C5–T1 nerve root injury

MOTOR LOSS		NERVE ROOT
Shoulder	Abductors and external rotators	C5
Elbow	Flexors	C5–6
Forearm	Pronators and supinators	C6
Wrist	Flexors and extensors	C6–7
Hand	Extrinsics	C7–8
	Intrinsics	T1
SENSORY LOSS		
Shoulder		C5
Hand	Thumb	C6
	Middle finger	C7
	Little finger	C8
AUTONOMIC LOSS		
Eye	Horner's syndrome	T1

PATHOLOGY

214. Nerve degeneration (Wallerian) and regeneration. When there is complete nerve division the distal segment undergoes Wallerian degeneration, i.e. the axons degenerate but the connective tissue sheaths remain open to accept any regenerating fibres from the proximal nerve end which may find their way into them.

214

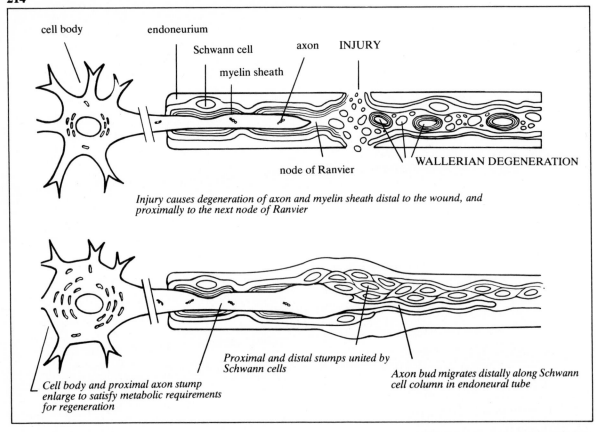

Injury causes degeneration of axon and myelin sheath distal to the wound, and proximally to the next node of Ranvier

Cell body and proximal axon stump enlarge to satisfy metabolic requirements for regeneration

Proximal and distal stumps united by Schwann cells

Axon bud migrates distally along Schwann cell column in endoneural tube

SENSORY SIGNS

215. Blisters in a patient with a divided median nerve. He was unable to feel heat and accidentally burnt himself with a cigarette.

215

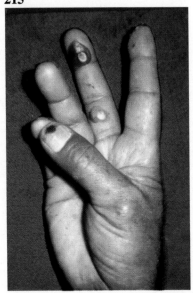

Augustus Volney Waller (1816–1870). English physiologist. 'Wallerian Degeneration' is the disintegrative process which occurs in the peripheral segment of a divided nerve.

216 & 217. Digital nerve division. Diminished dirt staining and sweating of the left thumb. Both digital nerves were divided by a knife.

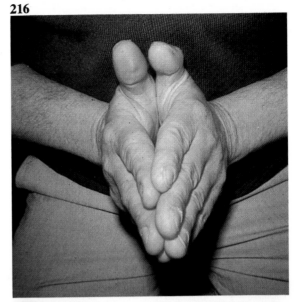

217. Digital nerve division. The same patient as in **216** undoing his shirt button with the dorsal aspect of his thumb – this had sensation from the uninjured digital branches of the radial nerve. He avoided using the numb pulp.

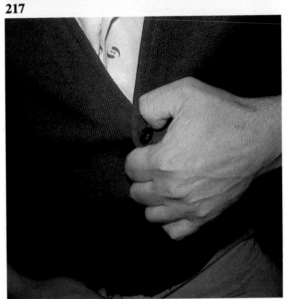

218. The pin-prick test for pain sensation. The patient's hand lies resting on a table. After the test is explained, he is blindfolded or asked to look the other way. He is then requested to identify and localise the site and nature of the sensory stimulus.

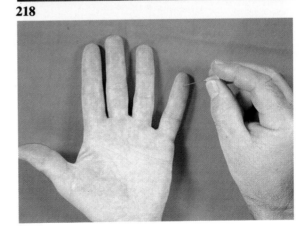

219. The cotton wool test for light touch.

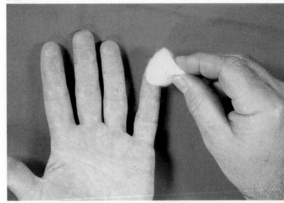

220. The pick-up test for stereognosis. The blind-folded patient is asked to pick up a variety of small objects, identify them and transfer them from the table top into a container. Note that this patient avoids using his right index and middle fingers which are numb from digital nerve injury.

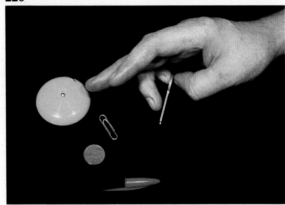

MOTOR SIGNS – as seen after ulnar nerve injury at the wrist. See **257–265**.

221. Wasting of the intrinsic muscles of the left thumb, and weakness of pinch.

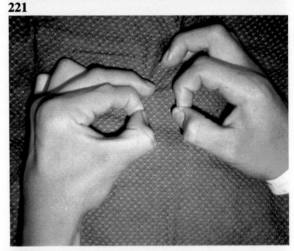

222. Wasting of the intrinsic muscles of the right hand and wasting of the right little finger in a six-year-old Chinese child.

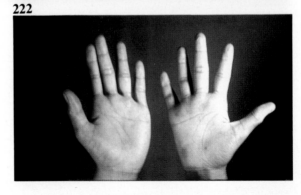

223. Early finger deformity – hyperextension of the metacarpo-phalangeal joint and flexion deformity of the proximal interphalangeal joints.

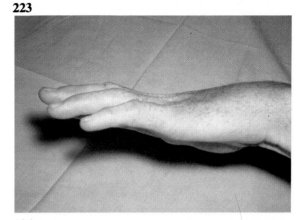

224. The ulnar claw hand. A later stage deformity from intrinsic muscle paralysis. The stippled area represents the associated sensory loss.

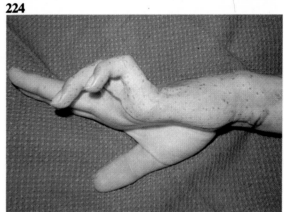

225. A reversible contracture. Passive flexion of the metacarpo-phalangeal joint allows full interphalangeal joint extension. This is the same patient as in **224**.

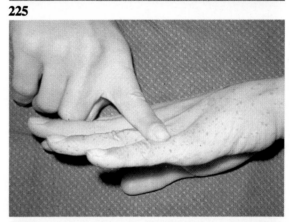

226. Fixed irreversible flexion deformity of the left ring and little fingers and associated intrinsic muscle wasting from long standing ulnar nerve palsy and secondary joint contracture.

AUTONOMIC SIGNS

In the early stages autonomic paralysis manifests as lack of sweating; later as vasomotor instability.

227. Lack of sweating, with dryness of the skin after nerve injury on the left index finger.

227

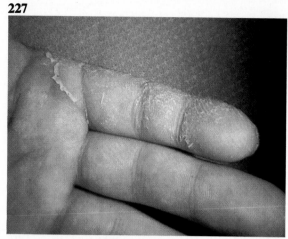

228a & 228b. Lack of normal skin wrinkling. Denervated skin remains smooth after immersion of the entire hand and forearm, in warm (40°C) water for 20 to 30 minutes. Normal wrinkling of the skin is seen in the thumb, index, and the ulnar side of the little finger. Lack of wrinkling is seen in the middle, ring, and radial side of the little finger. An atrophic (neuropathic) ulcer is seen on the middle finger.

The operative picture shows the division of the digital nerves.

This test was described by S. O'Riain in 1968. Following an observant mother's report of the return of wrinkling in her son's fingers in the bath following nerve repair, this phenomenon was studied on patients and found to be a useful test of nerve deficit and recovery.

229. The starch test. Digital nerve injury has resulted in lack of sweating on the ulnar side of the left thumb. This is shown by the absence of starch stain. The imprint of the normal right thumb is shown between the thumb and index finger.

230. Absent pigmentation and hair growth. This may occur in neuropathic skin. The patient has leprosy.

228a

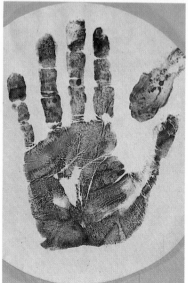

228b

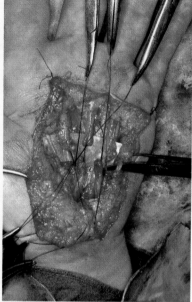

229

230

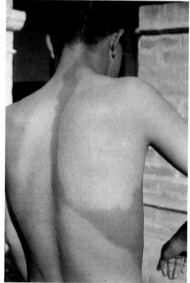

Séamus Ó'Riain. Contemporary Irish hand surgeon.

Miscellaneous signs

231

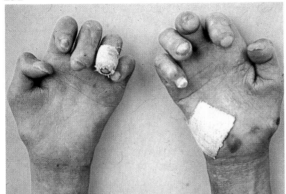

232

231. Gross neuropathy with trophic ulcers and contractures in advanced leprosy.

232. Hyperkeratosis on the gripping surfaces of an 'ulnar claw hand'. The middle, ring and little fingers are short and stubby, having been worn away by prolonged pressure. There is a trophic ulcer on the radial side of the hand. This 25-year-old Chinese male had leprosy.

233a & 233b. Neuropathic disintegration of the proximal interphalangeal joints of the index and middle fingers in a Chinese leper. These painful, swollen Charcot joints were mistreated as infection.

233a

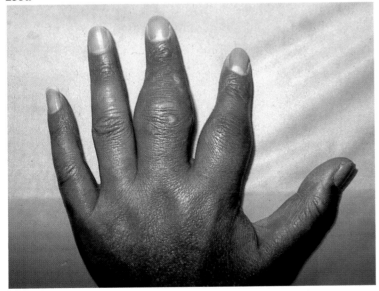

233b

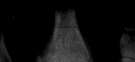

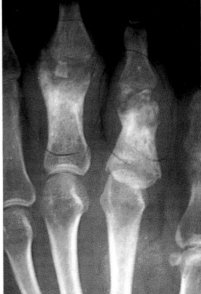

Variations of brachial plexus palsy

TYPE AND LEVEL OF LESION	MOTOR LOSS	SENSORY LOSS	AUTONOMIC LOSS
Upper – C5, 6 'Erb-Duchenne palsy' 'Porter's tip' position	Shoulder – abduction and external rotation Elbow – flexion Forearm – supination	C5 dermatome i.e. shoulder	—
Upper and mid – C5, 6, 7 ± 8	*As above plus* Forearm – supination and pronation Fingers – flexion (grip)	Nearly complete loss	—
Mid and low – C7, 8, T1 'Klumpke paralysis'	Hand – intrinsic and extrinsic paralysis 'Simian hand'	All except the thumb and index finger	'Horner's syndrome'
Complete – C5, 6, 7, 8, T1	Flail arm and hand	Complete loss	'Horner's syndrome'

Guillaume Duchenne (1806–1875). French physician from Boulogne. Devoted years of his life to the study of muscular action. *Erb-Duchenne paralysis* derives its name from the publication in 1877 by Erb, who quoted Duchenne's observation of similar paralyses in the newborn.

Wilhelm Heinrich Erb (1840–1921). German professor of neurology in Heidelburg. In 1873 he described the upper brachial (C5, 6) palsy. Duchenne had noticed this in the newborn in 1872, hence Erb-Duchenne palsy.

Specific patterns of peripheral nerve injury

INJURIES TO THE BRACHIAL PLEXUS

Each of these patients was aged between 19 and 22, and was involved in a motor cycle accident.

234 & 235. Right upper brachial plexus palsy, C5,6. 'Erb's palsy', 'Porter's tip position'. Paralysis of shoulder abduction and external rotation, elbow flexion and forearm supination. Sensory loss over the shoulder.

236–238. Right upper and mid brachial plexus palsy, C5,6,7±8. Paralysis of the shoulder and elbow as above, plus paralysis of the forearm supinators and pronators and finger flexors. The patient is unable to make a grip. Nearly complete sensory loss.

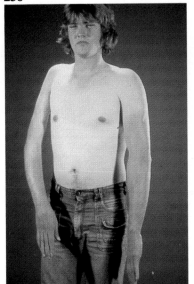 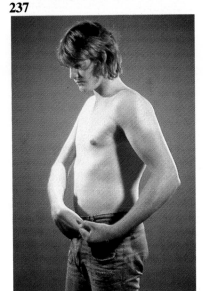 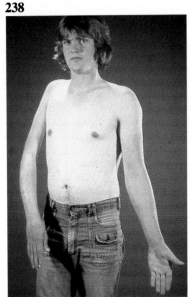

239

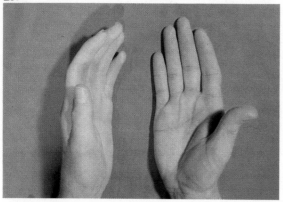

240

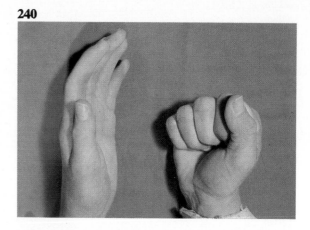

239 & 240. Left mid and low brachial plexus palsy, C7,8, T1. Klumpke's paralysis. Paralysis of the intrinsics (the 'Simian hand'), and extrinsics (the patient cannot make a fist or fully extend the fingers or thumb). Sensory loss over the ulnar three digits. In ulnar nerve palsy the median innervated thenar intrinsics are spared.

Madame Augusta Klumpke (1859–1927). Born in San Francisco, but later moved to Switzerland; the first woman extern, then intern, in Paris. Whilst a medical student in Paris, in 1885, she described the type of brachial palsy resulting from damage to the lower roots of the plexus.

241–243. Left complete brachial plexus palsy, C5,6,7,8, T1. Paralysis of all muscles of the upper limb. Complete loss of sensation. Horner's syndrome.

241

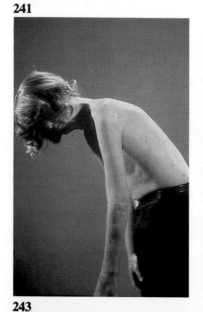

242

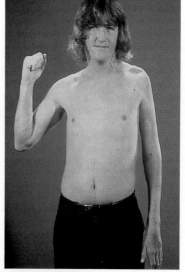

243

Johann Friedrich Horner (1831–1886). Professor of ophthalmology in Zürich, Switzerland. Described the syndrome of myosis, narrowing of the palpebral fissure and retraction of the globe.

Injury to radial, median, and ulnar nerves*

NERVE	LEVEL OF INJURY	MOTOR LOSS	SENSORY LOSS
Radial	Elbow	Wrist and digital M.C.P. joints – extension Thumb – extension and radial abduction	Radial ⅓, dorsum of hand
Posterior interosseous	Partial injury in upper forearm	Ulnar three fingers M.C.P. joints – extension	Nil
Median	Wrist	Thumb – loss of opposition (A.P.B., OPP.P., F.P.B.†) Index and middle fingers – 1st and 2nd lumbricals	Palmar surface radial 2½–3½ digits and dorsal aspect of their distal 2 phalanges
	Elbow	*As above plus* Forearm – pronation Wrist – radial deviation Fingers – flexion (F.P.L., all F.D.S., F.D.P. to index and middle fingers)	*As above*
Ulnar	Wrist	Thumb – adduction Fingers – abduction and adduction i.e. paralysis of all intrinsics except those supplied by median nerve	Palmar surface of ulnar 1½–2½ digits
Triple palsy – radial, median, and ulnar	Elbow	Hand – all extrinsics and intrinsics	The whole hand

*These are the most common findings – there are great variations in function of each nerve.

†The superficial head of FPB is supplied by the median and the deep head by the ulnar nerve.

INJURIES TO THE RADIAL, MEDIAN AND ULNAR NERVES

244

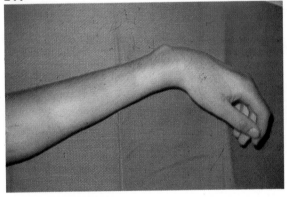

245

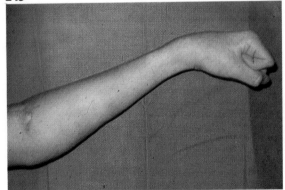

RADIAL NERVE

244. Radial nerve injury from laceration at the elbow. Inability to extend the wrist and the metacarpo-phalangeal joints of the fingers or thumb. 'Wrist drop'. The interphalangeal joints can be extended by the ulnar innervated intrinsics.

245. Normal flexion of fingers and thumb.

246. Incomplete division of the posterior interosseous nerve. Inability to extend the ulnar three fingers.

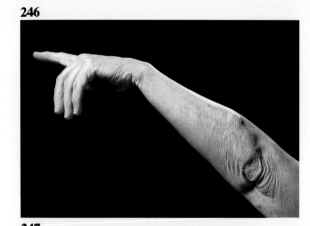

MEDIAN NERVE

247. Median nerve injury at the elbow. Loss of sensation over the median nerve distribution and loss of extrinsic flexor tendon function to the thumb and index finger.

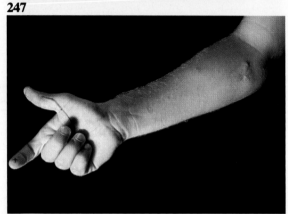

248. The 'Ochsner clasp test' showing loss of thumb and index finger flexion of the right hand.

249. Paralysed flexor digitorum superficialis function. Loss of independent flexion of proximal interphalangeal joint of middle finger. See **11**.

Albert John Ochsner (1858–1925). American surgeon.

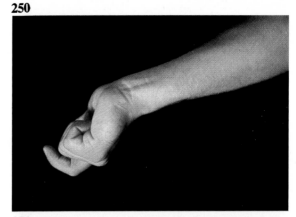

250. Normal function of the flexor digitorum profundus of the ring and little fingers. These are supplied by the ulnar nerve. Middle finger flexion is present, but weak.

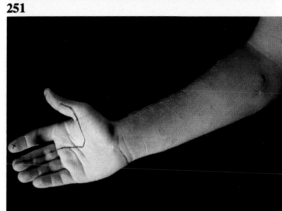

251. Sensory loss. Dryness and blistering of the skin over the thumb, index and half of the middle finger; in this patient the ulnar nerve supplies the ulnar 2½ digits.

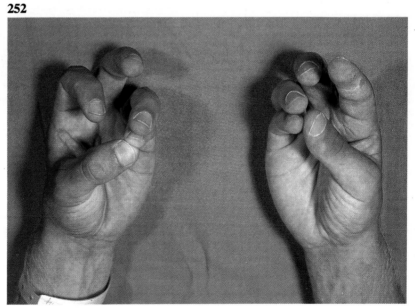

252. Failure of true opposition due to palsy of the abductor pollicis brevis, opponens pollicis and flexor pollicis brevis, i.e. the right thumb cannot align with the nail of the ring finger.

253

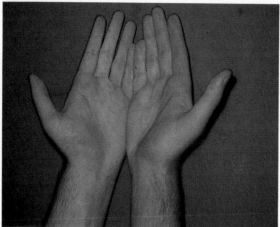

253–255. Median nerve injury at the elbow.
Wasting of the thenar branch of the median nerve. Wasting of the thumb, index, middle, and ring finger from disuse (from poor sensation and extrinsic flexor tendon function).

Wasting and paralysis of abductor pollicis brevis (**254, 255**).

254

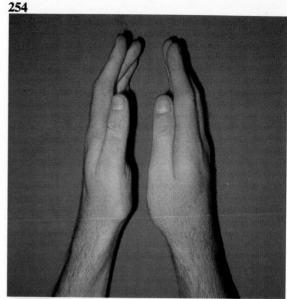

255

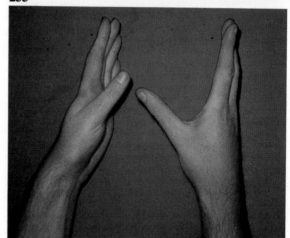

256. Median nerve injury at the wrist. Loss of sensation of the radial 2½ digits (the ulnar nerve in this patient also supplied the ulnar 2½ digits). Loss of thenar abduction and opposition. There is normal function of flexor profundus of the index finger and flexor pollicis longus because these muscles are supplied by the median nerve proximal to the injury.

256

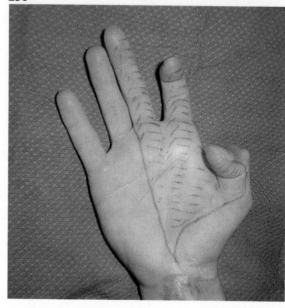

ULNAR NERVE

257. Ulnar nerve injury at the wrist. Abduction of the little finger. This is the first sign to appear and the last to go (after regeneration of the nerve). It is due to the unopposed action of the radially innervated extensor muscles, which also abduct the little finger.

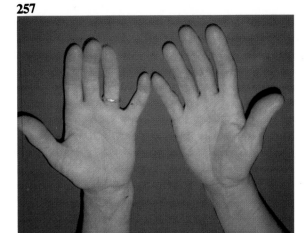

258. Ulnar nerve injury at the wrist. Lateral view showing intrinsic paralysis, i.e. inability to flex the metacarpo-phalangeal joints and inability to extend the interphalangeal joints.

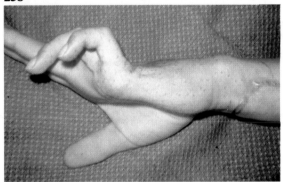

259. Loss of sensation and early flexion deformity of the ring and little fingers. The 'ulnar claw' hand.

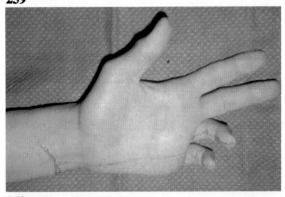

260. Full finger flexion is present in ulnar nerve palsy at the wrist because the extrinsic flexors are innervated by the ulnar nerve proximal to the injury. The same patient is seen in **261, 262,** and **265**.

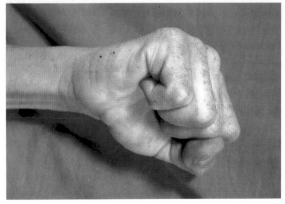

261

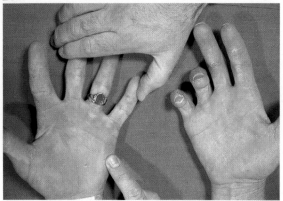

261. Testing intrinsic abduction of the fingers against resistance. The right hand shows wasting and claw deformity from ulnar nerve palsy.

262. The 'card test' for adduction. The patient is asked to grip a card between his extended fingers. The right index and middle fingers have no adduction power.

263. Testing for abductor pollicis. Froment's sign. Ask the patient to grasp a piece of paper firmly between the thumb and index fingers of each hand. Then ask him to pull the hands away from each other without releasing the grip. When there is paralysis of abductor pollicis, supplied by the ulnar nerve, there is compensatory over-reaction of flexor pollicis longus which flexes the interphalangeal joint. This patient had division of his right ulnar nerve at the wrist.

264. Wasting of the first web in ulnar nerve palsy (paralysis of the first dorsal interosseous and adductor pollicis muscles).

265. Weak pincer function of the right hand.

262

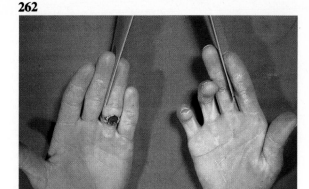

263

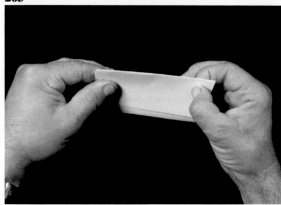

264

265

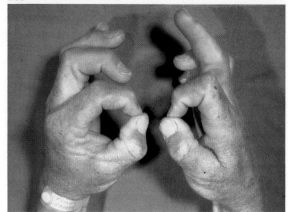

Jules Froment (1878–1946). French neurologist and professor of medicine. Described the 'newspaper draw sign', illustrating the peculiar type of pinch with the thumb when the adductor pollicis is paralysed.

COMBINED RADIAL, MEDIAN AND ULNAR NERVE

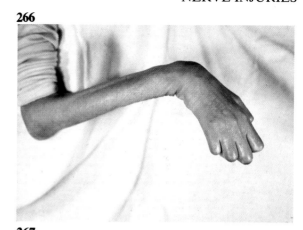

266. Triple palsy of the radial, median and ulnar nerves. This patient had leprosy.

267 & 268. Combined median and ulnar nerve palsy at the wrist. Paralysis of all intrinsic muscles.

DIGITAL NERVE

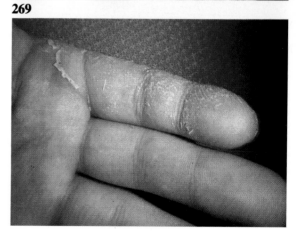

269. Digital nerve injury. Area of numbness after division of the digital nerve. The skin is dry from lack of sweating.

Vascular injuries

Acute vascular injuries

As a result of direct or indirect injury the following vascular changes may occur – spasm, thrombosis or division. Penetration of both artery and vein may produce an arteriovenus fistula. Penetration of an artery wall may produce a false aneurism.

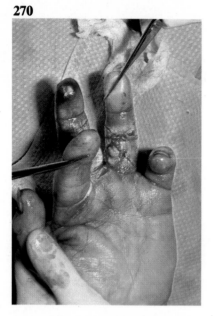

270. Crush injury. Gangrene of the left middle finger and ischaemia of the adjacent fingertip from crush injury to the base of the fingers.

The operative findings included contusion, spasm and thrombosis of the digital arteries and veins.

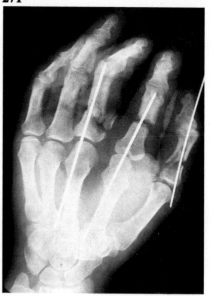

271. Extensive fractures of the proximal phalanges, subsequently stabilised by wires.

Martin Kirschner (1879–1942). Austrian surgeon who first described new small calibre steel wires to fix bony fragments in place or to immobilize a joint.

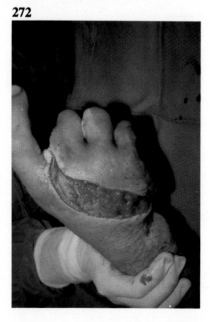

272. Acute ischaemia in a hand caught between rollers. Dorsal view. Oedema and congestion of the avulsed flap due to inadequate venous return.

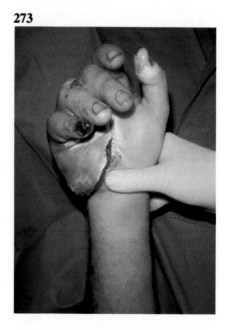

273. Ischaemia of the flap and gangrene of the little finger. Palmar view.

274. Gangrene of the thumb 24 hours after a mangling injury which twisted both neurovascular bundles – the 'life lines' of the digit.

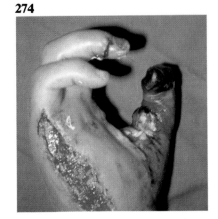

Gangrene. In Greek gangrene means to gnaw or feed upon. The term was used by Galen (Claudius Galenus, Greek physician, c. 129–199 AD) for an eating sore, ending in mortification.

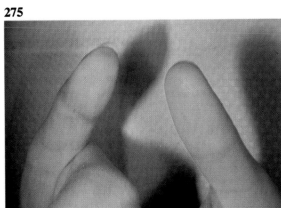

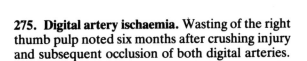

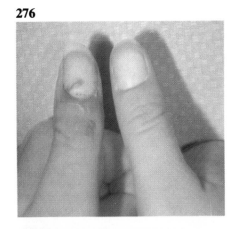

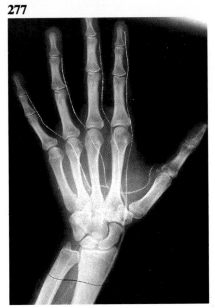

275. Digital artery ischaemia. Wasting of the right thumb pulp noted six months after crushing injury and subsequent occlusion of both digital arteries.

276. Wasting and scarring of the thumb and nail, dorsal view.

277. Angiogram showing absence of vascular filling distal to the metacarpo-phalangeal joint of the thumb.

Chronic vascular injuries

278

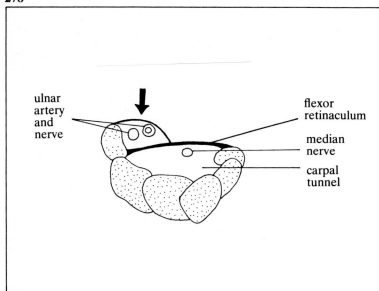

279

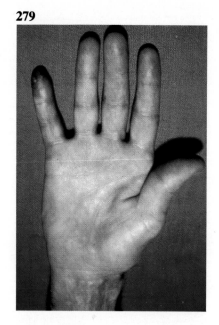

278. 'Hand hammer syndrome'. Repeated blows to the hypothenar eminence can produce clinical features of ulnar artery and nerve compression.

279. 'Hand hammer syndrome'. This 60-year-old carpenter used his hypothenar eminence as a hammer. His little finger became cold, blue and numb. The physical findings of diminished blood supply included diminished capillary filling, absent digital pulses and tenderness over the ulnar artery at the wrist.

280. 'Digital hammer syndrome'. Chronic ischaemic ulcer of the right middle finger from repeated striking of the base of the finger against a heavy spanner.

281. Poor perfusion of the digital blood vessels to the right middle finger.

280

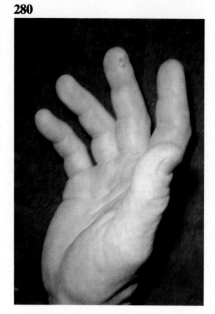

281

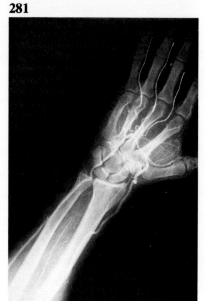

282. Raynaud's phenomenon in a 49-year-old man whose working life was spent using a jack hammer. Note the blueness of the fingertips of his right hand.

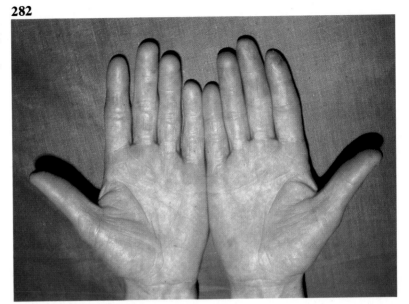

282

Traumatic aneurysm and arteriovenous fistula

A traumatic arteriovenous fistula should be suspected if a painful warm swelling develops near a neurovascular bundle after penetrating injury. A bruit may be heard or felt, and can be eliminated by proximal compression of the artery. Varices may develop distally.

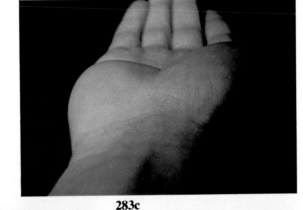

283a

283a. Traumatic (false) aneurysm of the ulnar artery presenting as a painless pulsatile swelling of the hypothenar eminence.

283b. Arteriogram of the traumatic aneurysm.

283c. Operative findings of the traumatic aneurysm.

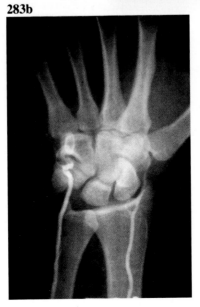

283b

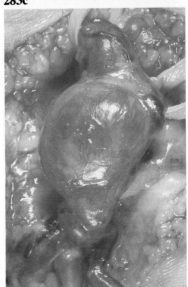

283c

Iatrogenic and self-inflicted vascular injuries

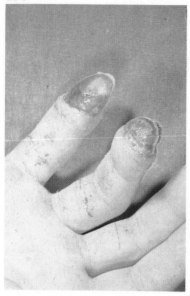

284. Ischaemia of the fingertips complicating a digital nerve block. The local anaesthetic which was used contained adrenalin.

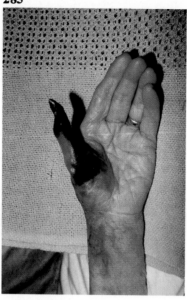

285. Ischaemia complicating radial artery cannulation.

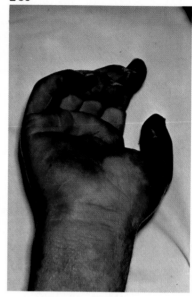

286. Ischaemia of the hand from intra-arterial injection of barbiturates.

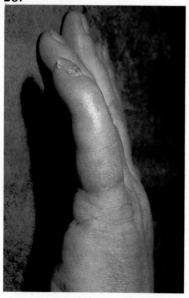

287. Ischaemia from a rubberband tourniquet.

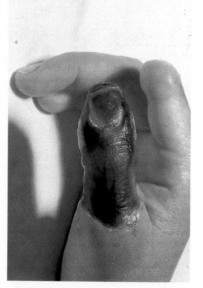

288. Gangrene of the thumb from prolonged application of a tourniquet around the base of the thumb.

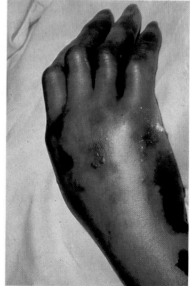

289. Severe venous congestion from a tight plaster at the elbow.

Miscellaneous

290. Chronic ischaemia from an overdose of x-rays. This 35-year-old, as a child, had been given x-ray treatment for warts. He developed ischaemic ulcers necessitating amputation of the index and ring fingers. He now has pain after prolonged use of his wasted hand.

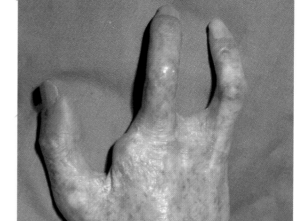

290

291. Chronic ischaemia manifesting as Volkmann's ischaemic contracture. Severe impairment in forearm of circulation following crush injury. Ischaemic fibrosis of the forearm flexors prevents extension of the fingers.

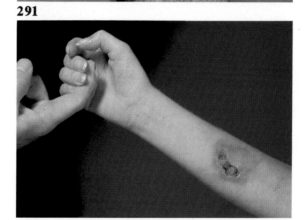

291

292. Spontaneous haemorrhage in a patient on anticoagulants.

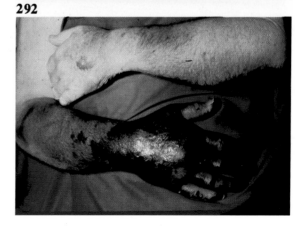

292

Richard Von Volkmann (1830–1889). One of the best known German surgeons of the mid-19th century described ischaemic contracture in 1881.

Important types of joint injury seen in the hand

JOINT	MECHANISM OF INJURY	TYPE OF DISLOCATION	SIGNIFICANCE
Proximal interphalangeal (PIP)	Hyperextension, crush or compression		PIP joint stiffness
Metacarpo-phalangeal (thumb) Rupture of ulnar collateral ligament (with or without fracture)	Angulation	*dorsal view of left thumb* ulnar collateral ligament	Instability, weak pinch
Carpo-metacarpal joint of thumb (Bennett's fracture)	Axial compression	radial and proximal dislocation of meta-carpal *palmar view of left thumb* trapezium	Fracture dislocation of the key joint of the thumb
Radiocarpal and intercarpal	Hyperextension of wrist	*palmar* *dorsal* capitate lunate median nerve normal perilunar dislocation semilunar dislocation	Median nerve compression, wrist joint stiffness

Edward Halloran Bennett (1837–1907). Irish anatomist and surgeon. Described the intra-articular fracture at the base of the first metacarpal bone in 1882. Known as Bennett's fracture or Boxer's fracture.

Skeletal injuries

293

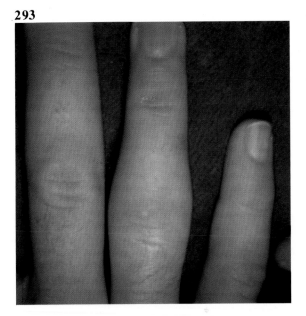

294

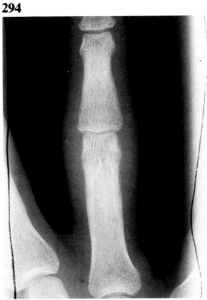

Joint injuries

293. Sprain of the proximal interphalangeal joint of the right ring finger. Pain, fusiform swelling and stiffness may persist for months.

294. Soft tissue swelling, x-ray. There are no signs of fracture, dislocation nor ligament rupture.

295

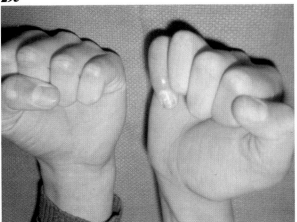

296

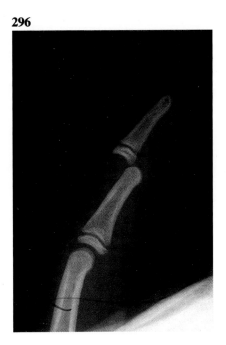

295. Dislocation of the distal interphalangeal joint of the right little finger. There was full flexion of all joints except the dislocated joint.

296. X-ray. The distal segment generally dislocates dorsally.

297a, b. Joint injury with ligament rupture. (Ulnar collateral ligament of the thumb). After angulation injury, the metacarpophalangeal joint of the thumb became painful, swollen and stiff. The initial diagnosis was a 'sprain'.

Rupture of the ulnar collateral ligament was diagnosed by first anaesthetising the thumb and then applying lateral stress to the extended metacarpo-phalangeal joint.

297a 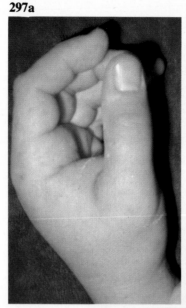 297b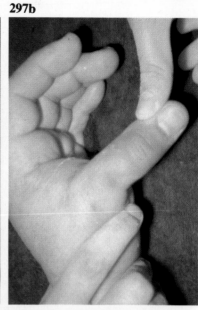

298. X-ray, stress view. Abnormal radial deviation of the left thumb. Right thumb normal.

298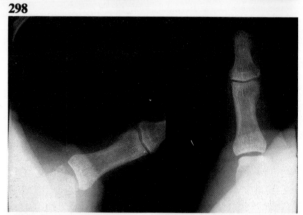

299. Fracture separation of the bony attachment of the ulnar collateral ligament after a similar injury (arrow).

300. Minute fracture at the base of the middle phalanx (arrow). There are no important ligament or tendon attachments to this fragment. After the initial traumatic inflammatory response there are usually no residual clinical signs.

299 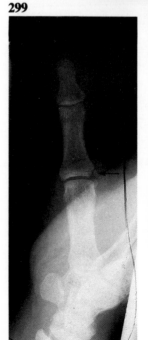 300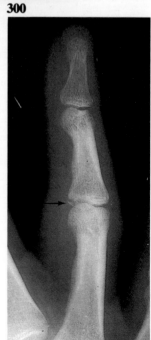

301. Intra-articular fracture of the proximal inter-phalangeal joint with secondary joint dislocation. The medial extensor tendon pulls the dorsal fragment dorsally and the flexor tendons pull the volar fragment and the proximal interphalangeal joint into volar dislocation.

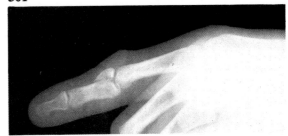

302. Anterior dislocation of the lunate; semi-lunar dislocation. This patient, after a severe hyper-extension injury, complained of pain, swelling, and paraesthesia in the median nerve distribution. On a true antero-posterior view the normal lunate is surrounded by a clear 1mm zone, representing articular cartilage. See **69**. Loss of, or widening of, this zone suggests the dislocation that is best seen on a lateral x-ray.

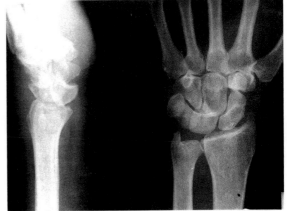

303. Peri-lunar dislocation. The lunate lies in the correct position. The capitate and carpus are dislocated posteriorly. There is an associated fracture of the ulnar styloid.

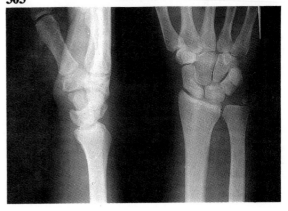

Important types of fracture seen in the hand

BONE	MECHANISM OF INJURY	TYPE OF FRACTURE	SIGNIFICANCE
Distal phalanx	Crush	tuft — comminuted — longitudinal — transverse	These fractures are of secondary importance to the associated soft tissue injuries
	Traction, avulsion	terminal extensor tendon — flexor digitorum profundus — flexor digitorum profundus	Joint deformity from muscle – tendon imbalance
Metacarpal, proximal, or middle phalanx	Axial compression	metacarpal neck fracture	These fractures tend to angulation deformity, leading to secondary contracture of the proximal interphalangeal joint
	Direct blow	intrinsic pull	
	Twist or torque	spiral fracture	These fractures tend to *rotational* and shortening deformity
Scaphoid	Dorsi flexion and radial deviation	1. tubercle 2. waist 3. proximal pole	1. Good healing 2. Fair healing 3. Poor healing

Fractures

The common types of hand fractures and their significance are shown in the table.

Fracture of one of the twenty-two bones comprising the skeletal frame can so alter the balance of tendon forces as to produce severe disorder of hand function.

304. Muscle tendon forces causing angulation of a fractured proximal phalanx. Extensor digitorum and the lumbrical pull the head of the proximal phalanx dorsally, the interossei pulls the base of the proximal phalanx volarly. Angulation then becomes self-perpetuating.

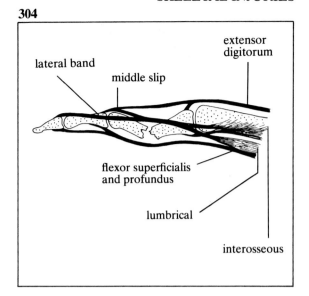

304

lateral band

middle slip

extensor digitorum

flexor superficialis and profundus

lumbrical

interosseous

305 & 306. Transverse fracture with angulation deformity – caused by the pull of the intrinsic and extrinsic flexor and extensor forces. The shorter the distal fragment the greater the angulation.

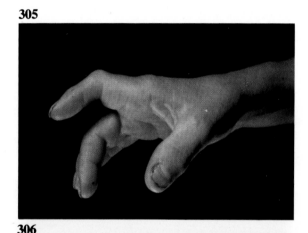

305

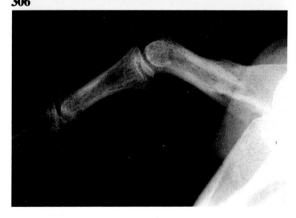

306

307

308

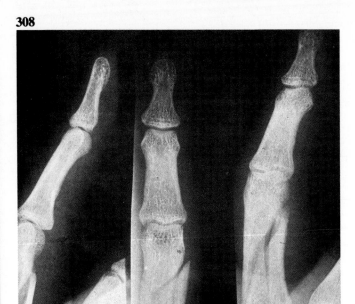

307 & 308. Spiral fracture of the proximal phalanx (ring finger) with rotational deformity and shortening – the pull of the tendon forces causes the two fracture fragments to slide on each other. All fingertips should point to the tubercle of the scaphoid.

Displaced fractures can also cause nerve injury e.g. ulnar nerve palsy developed after closed fracture of the ulna. (See **209**).

Fingertip injuries

These may be open or closed. The extent of soft tissue damage can be diagnosed by knowing the exact mechanism of injury and by inspection. An x-ray is necessary to diagnose the skeletal damage.

Open

309. The various levels of fingertip injury. Note the relationship of the levels of pulp, nail and phalanx injury.

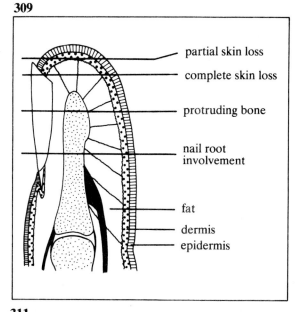

partial skin loss

complete skin loss

protruding bone

nail root involvement

fat

dermis

epidermis

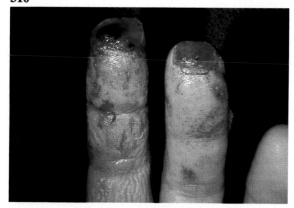

310. Partial thickness skin loss. Bleeding from the dermis. Subcutaneous fat not visible.

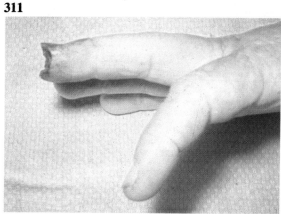

311. Complete skin loss. Subcutaneous fat visible.

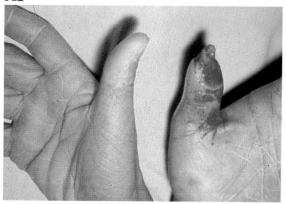

312. Complete skin loss. Bony phalanx protruding.

313. Incomplete avulsion of pulp. The flap is viable because there is an intact neurovascular bundle.

314. Complete avulsion of pulp tissue.

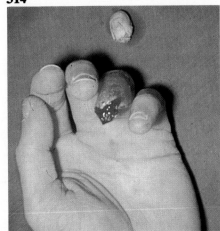

314

Closed

These may involve the pulp, nail apparatus, flexor or extensor tendons, distal phalanx or distal interphalangeal joint.

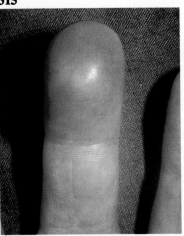

315

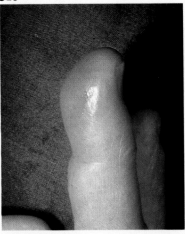

316

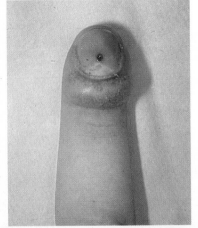

317

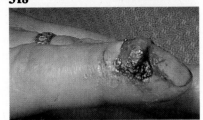

318

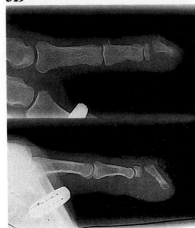

319

315 & 316. Crush injury with distal pulp haematoma. This patient jammed his finger in a car door. He had very severe throbbing pain, relieved only by surgical decompression.

317. An extensive subungual haematoma. Blood is seen distending the nail folds. The nail has been punctured to relieve tension.

318. Dislocation of the nail apparatus and fracture of the distal phalanx following crush injury to the fingertip.

319. X-ray. Flexor digitorum profundus is attached to and is displacing the distal fragment. The extensor mechanism is attached to the proximal fragment.

320

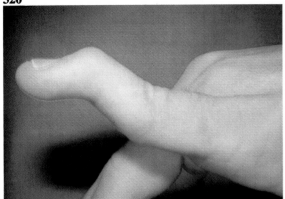

321

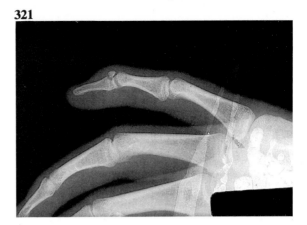

320 & 321. Mallet finger with closed rupture of the terminal extensor tendon following a crush injury.

Foreign bodies

Foreign bodies may present as:
(1) Tumour (cyst)
(2) Infection (sinus)
(3) Deep damage (tendon or nerve)

322–324. Tumour on the back of the hand. Laceration by glass three years ago. X-ray shows the glass foreign body; the operative finding is glass walled off by granulomatous tissue.

322

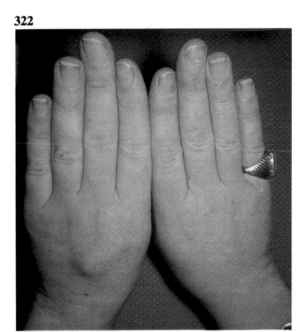

323

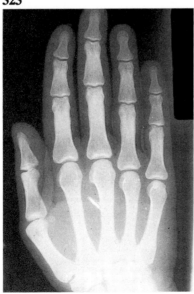

324

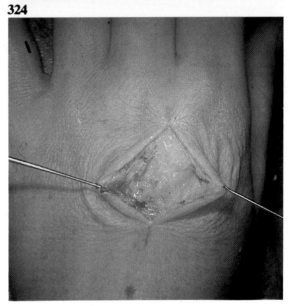

325. Cystic tumour in the proximal pulp of the ring finger appearing three months after penetration by a fish scale.

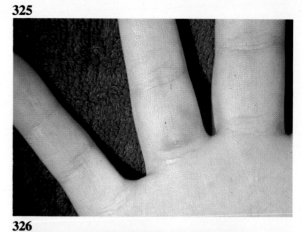

326. X-ray of the fish scale, seen only on the lateral x-ray.

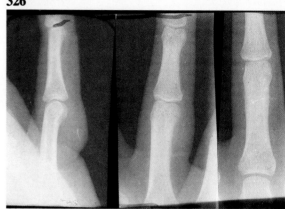

327. Operative specimen of a foreign body reaction to a wooden splinter.

328. Foreign body reaction to catgut suture in a tendon.

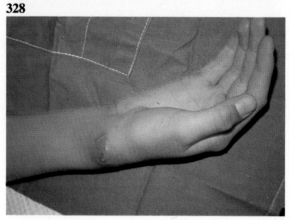

329. Steel pin foreign body. Very little tissue reaction.

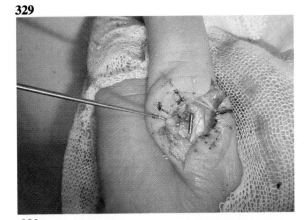

330. X-ray of the steel pin. This foreign body was found by marking the skin with crossed pins.

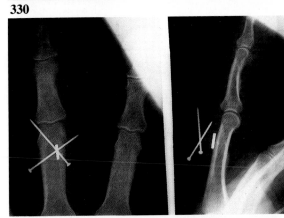

331. X-ray diagnosis. Three views were needed for precise location.

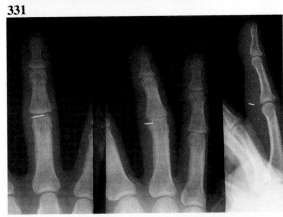

332. Operative findings.

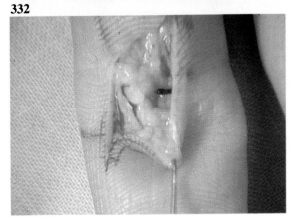

333

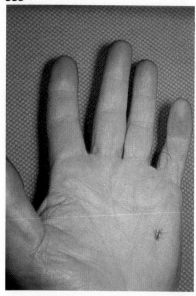

333. Sinus due to a glass foreign body. This 55-year-old male noticed a persistent sore on his hand. A whisky bottle had exploded in his hand eight years before.

335

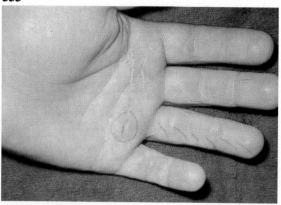

335. Nerve division with numbness of the ulnar aspect of the left ring finger from a glass fragment foreign body embedded in the distal palm six months before.

336. An unusual foreign body. Accidental penetration by a ramset nail. The patient had no symptoms.

337. Fish hook embedded in the base of the thumb. The patient demonstrates a similar hook.

Pyogenic granuloma due to a retained thorn foreign body. See **463**.

334

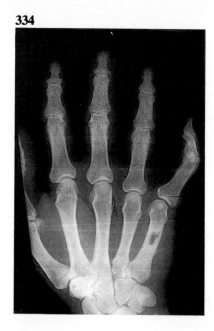

334. X-ray showing glass foreign body and an infective granuloma in the 5th metacarpal.

336

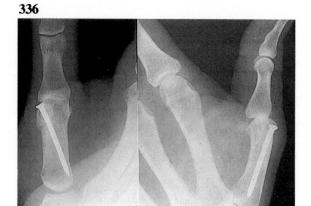

337

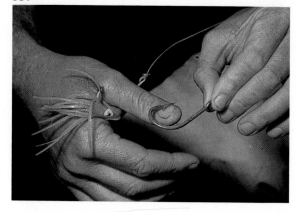

Burns

Most burns of the hand are caused by heat. Other causes include friction, electricity, chemicals irradiation, and cold.

Exposure burns occur when the hand is used to protect the face and the rest of the body.

Contact burns occur when the hand grasps hot objects.

Pathological classification. There are three pathological types, based on the depth of burn injury.

(*a*) Superficial burns. These are painful because the nerve endings remain uninjured. They present as erythema (1st degree) or blistering (2nd degree).

(*b*) In deep burns (3rd degree) there is loss of all epithelium and superficial nerve endings. These burns are therefore painless and present as greyish or charred indurated areas.

(*c*) In many burns there is a *mixture* of both superficial and deep tissue loss.

338. Three depths of burn.

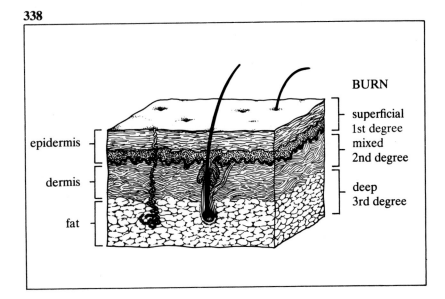

338

epidermis

dermis

fat

BURN

superficial
1st degree

mixed
2nd degree

deep
3rd degree

Superficial burns

339. Superficial burns from brief exposure to flame. There is painful erythema.

339

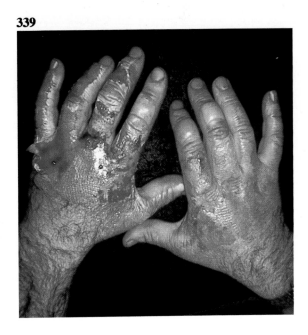

340

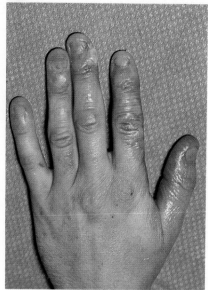

341

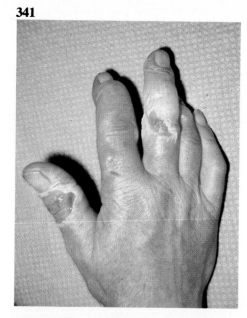

340. Superficial exposure burns from an electric flash.

341. Superficial burns from the steam of a boiling car radiator.

342

343

342 & 343. Superficial contact burns from accidental immersion of the hand in hot fat. Painful blisters.

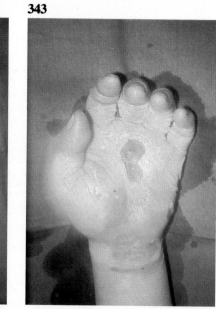

344. The extent of erythema seen after removal of the blisters.

344

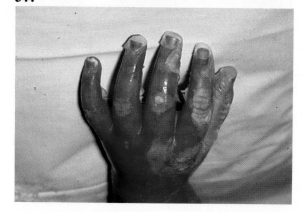

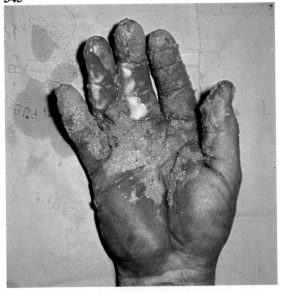

345. Superficial contact burns from tar. Tar only causes superficial contact burns because it cools quickly.

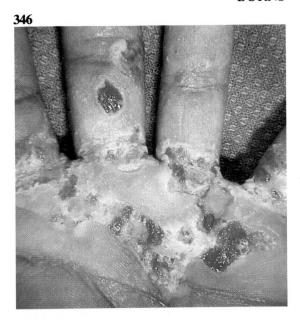

346. Superficial contact burns from lime.

Deep burns

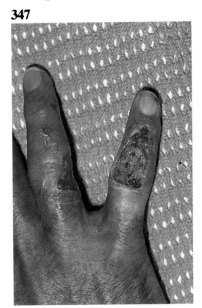

347. Deep contact burns from lysol. Loss of skin and extensor tendon.

348. Deep contact burn from grasping a hot roller.

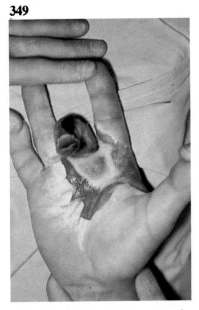

349. Deep burn from grasping high voltage electric wires. Tendon involvement causes a flexion contracture. Electrical burns are associated with deep tissue damage and they simulate a crush injury more than a thermal burn.

350

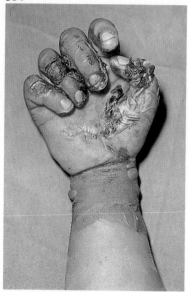

351

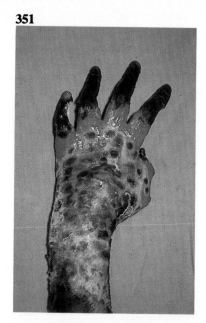

350. Electrocution. This 12-year-old aboriginal boy grasped high voltage electric wires. The blackened, brown dried skin is characteristic, but even more typical is the way in which the surface layers of the skin have been thrown back from the underlying tissues, due probably to the effect of the heat generated in the tissue fluids. He subsequently had amputation of the right hand. There was gross destruction of deep tissues.

351. Deep burns and gangrene. The fingers were injured by a combination of heat and friction in laundry rollers.

353

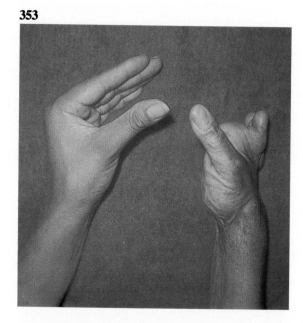

352

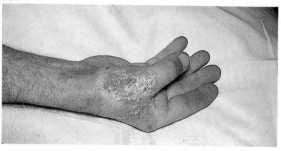

352. Radiation necrosis. See also **290** and **806**.

353. The aftermath of burns. Burn contractures in a 21-year-old Botswana female whose hand was burnt by flame when she was a child.

Miscellaneous

Self-inflicted

The hand is the most common place for a self-inflicted or self-perpetuated injury as it is the most convenient place for the patient to achieve the injury and to show its results.

The diagnosis is usually obvious if one is aware of the patient's motive which is usually monetary or emotional gain.

There are also various types of artefact and functional lesions.

354

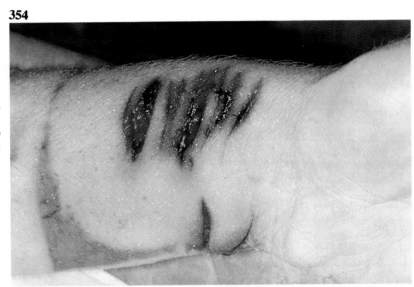

354. Attempted suicidal lacerations at the wrist. There is much bruising around these wounds because they were inflicted with a blunt knife. If the knife had been sharper the bruising would have been less. The occurrence of the cuts in a group is more suggestive of accident than murderous assault.

355

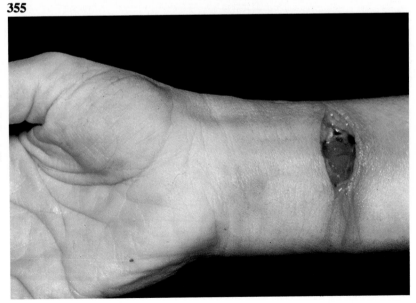

355. Typical suicidal cuts on the wrist. There are several tentative, superficial incisions before the final deep incision. In this case the muscle and tendon were incised, but no major artery was divided. Attempted suicide by wrist cutting is usually unsuccessful. However, this woman also cut her throat and died as a result of bleeding.

356

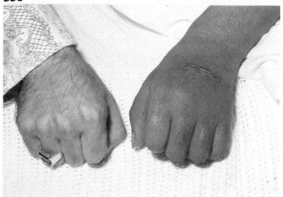

356. Self-inflicted oedema from repeated application of a piece of string around the wrist. Note the transverse pressure scar over the dorsum of the left hand.

357

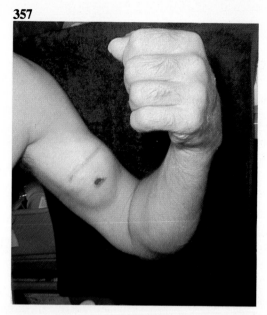

357. Self-inflicted constriction ring around the upper arm in the same patient. A pressure sore from a previous tourniquet is seen below the constriction.

358

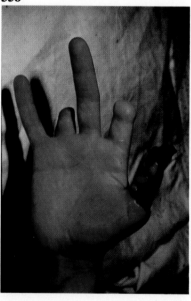

358. Self-inflicted gangrene. The patient tied a piece of cotton around the base of her thumb. She had previously achieved a similar result on the index and ring fingers.

359

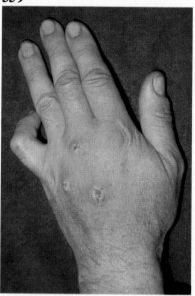

359. Self-inflicted sores on the dorsal aspect of the left hand. This patient, who initially had a mild sprain of his wrist, used a tooth pick to continually prick his hand to gain higher compensation payments.

Note the hysterical flexor spasm of the little finger, and the extensor spasm of the other fingers and thumb.

360

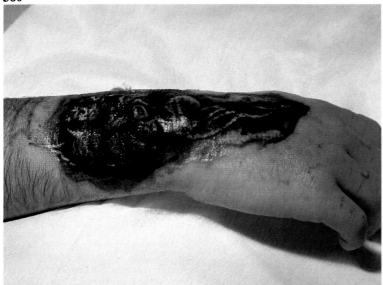

361

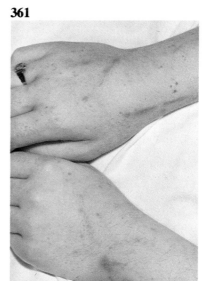

360. Self-inflicted burns. This patient used sodium hydroxide in an attempt to remove tattoos.

361. Drug injection and tattoos in a heroin addict. Repeated venepuncture led to tattooing of the skin over the veins of the hand and wrist.

362

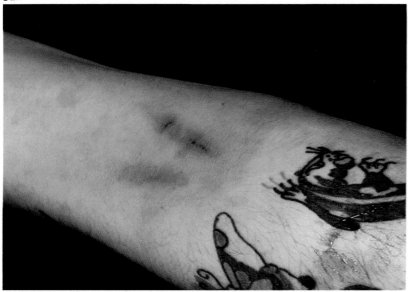

362. The tattooed arm of an addict. Note the purple linear marks with puncture holes in them in front of the elbow. These are veins that have been thrombosed because various drugs and fluids have been injected into them by the individual himself.

363a, b. Tattoo marks on the hand of a drug addict. Such curious tattoo marks often accompany the other stigmata of drug addiction that are illustrated in the previous figures.

363a

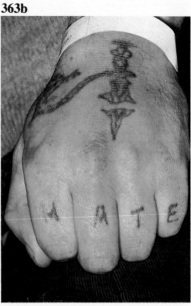
363b

Functional

364. Hysterical adduction contracture of the thumb occurring one month after removal of a thorn from the pulp. The index finger was previously injured, and amputated.

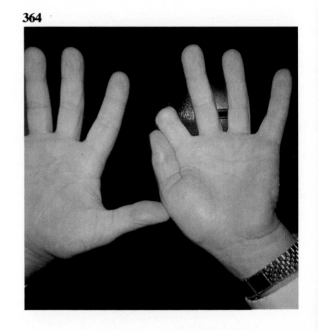
364

Bites

Bite wounds. The hand is liable to bite from animals, snakes, spiders, insects and humans.

Human bites over the knuckle almost always penetrate the metacarpo-phalangeal joint. See Chapter 4, Infections.

Snake bites. Poisonous snakes leave two puncture wounds. There are four types of clinical manifestation:

(a) local – pain, redness and swelling.

(b) general – nausea, vomiting and collapse.

(c) haemotoxic – haemolysis, local haemorrhage and bleeding.

(d) neurotoxic – respiratory failure.

Non-poisonous snakes leave semi-circular rows of tooth marks.

365. Puncture wounds of the finger from a black snake bite (poisonous).

366. Brown snake bite with signs of haemolysis in the upper arm.

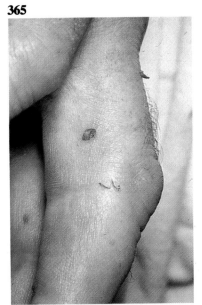

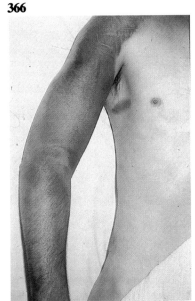

367. Frost 'bite'. Splinter haemorrhages in the recovery phase.

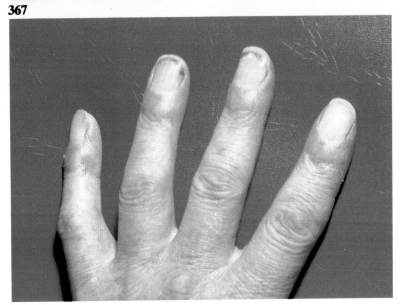

4. Infections

So intricate is the structure of the hand that even minor degrees of inflammatory oedema or necrosis can cause major functional disorder.

Though infections are found in any tissue, they occur most commonly around the nail fold and the distal pulp. Here puncture wounds are frequent and the blood supply is easily obstructed.

Hand infections should make one suspect that the patient has diminished *general* resistance from diabetes mellitus, steroid therapy, peripheral vascular disease or blood dyscrasias.

Common causes of *local* diminished resistance to infection include the presence of a foreign body and local ischaemia. Bites and puncture wounds are often complicated by infection.

368

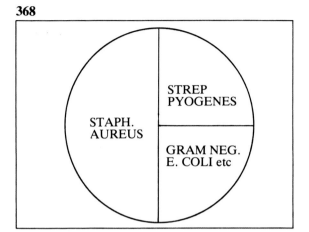

368. Approximate incidence of organisms cultured from hand infections.

369

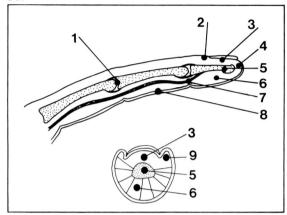

369. Sites of finger and fingertip infections. (1) septic arthritis, (2) eponychia, (3) subonychia, (4) apical abscess, (5) osteomyelitis, (6) distal pulp space infection, (7) flexor tenosynovitis, (8) skin abscess, (9) paronychia.

370

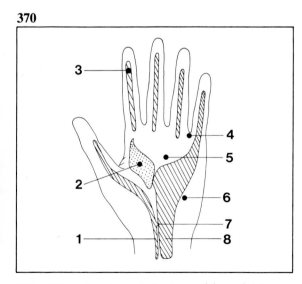

370. Sites of palmar infections. (1) radial bursa, (2) thenar space infection, (3) flexor tendon sheath infection, (4) web space infection, (5) mid palmar space infection, (6) hypothenar space infection, (7) communication, (8) ulnar bursa; also the site of flexor tenosynovitis.

371

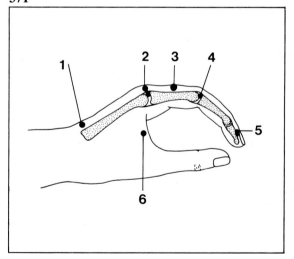

371. Sites of dorsal infections. (1) subcutaneous and subaponeurotic space infection, (2) bite wound infection and septic arthritis, (3) carbuncle, (4) septic arthritis, (5) paronychia, (6) thenar web infection.

Clinical features

Persistent pain is the characteristic symptom of a hand infection. The pain is throbbing in nature and if there has been one sleepless night an abscess is probably present.

Clinical features of hand infections

Symptoms	Signs
Throbbing pain	Dorsal swelling
	Restricted movement
	Local tenderness
	Fever

Investigations

(1) X-ray – for foreign body, bone and joint change.

(2) Full blood count – for blood dyscrasias and leukocyte response.

(3) Urinalysis and blood sugar – for diabetes mellitus.

(4) Serum uric acid – for gout.

372. Dorsal swelling. Though the site of infection is usually on the palmar side, maximal swelling will always occur in the loose dorsal subcutaneous tissue. This patient has a palmar space infection.

372

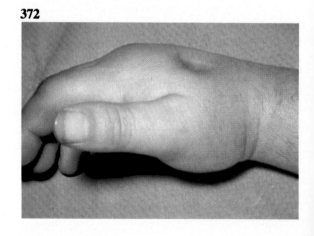

Cellulitis is a generalised infection. Patients usually present with a 12–24 hour history of throbbing pain. The onset is acute, and the patient appears ill and has a fever.

373 & 374. Streptococcal cellulitis of the right middle finger with lymphangitis of the hand and forearm. The draining lymph nodes are initially tender, and later enlarged.

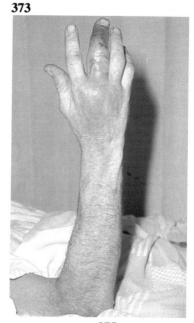
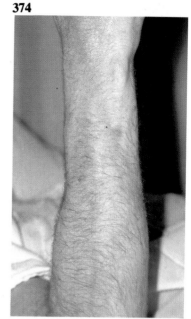

375. Streptococcal cellulitis of the left middle finger. The dorsum of the hand is swollen.

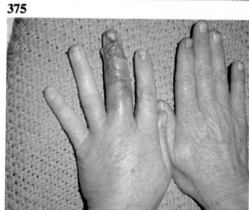

An **abscess** (i.e. suppuration) is localised infection with pus formation, usually caused by staphylococci. The patients frequently present with a 3–5 day history of throbbing pain with one or more sleepless nights.

376. Paronychial abscess. The end result of cellulitis.

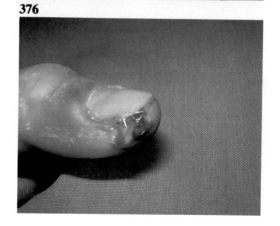

Skin

377. Skin abscesses, pyoderma, intracuticular abscess. An abscess may start in the skin as an infected blister. It may penetrate through the dermis to the subcutaneous space or through the palmar fascia into the palmar space and so form a collar stud abscess. Alternatively, a deep abscess may penetrate superficially through the palmar fascia or through the dermis and present as a subcuticular abscess.

377

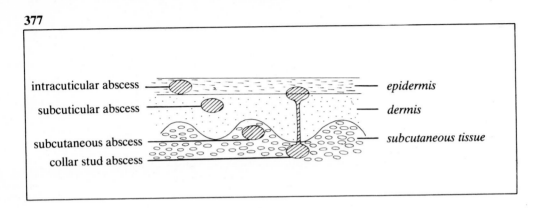

A **furuncle**, boil or pustule begins as a staphylococcal infection of a hair follicle, usually on the dorsum of the hand or over the proximal phalanx.

378

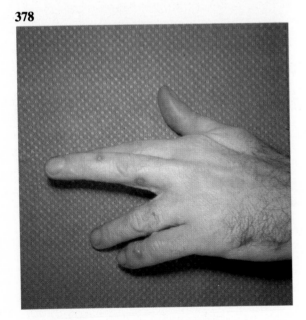

378. Multiple furuncles in a diabetic.

379

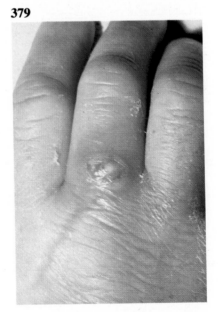

379. A subcuticular abscess which developed from a furuncle. There is a larger inflammatory reaction.

A **'collar stud abscess'** should always be suspected if there is tissue reaction out of all proportion to the size of the superficial component of the abscess.

380. Collar stud abscess resulting from stabbing of the thenar crease with an indelible pencil. The deep component of this abscess was in the mid-palmar space which became tender and swollen. The middle finger is flexed because of involvement of its tendon sheath.

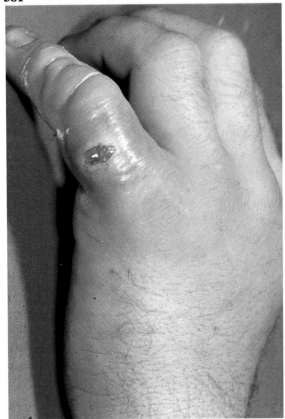

380

381

A **carbuncle** is infective gangrene of the skin and subcutaneous tissue. It begins as a furuncle but then extends and develops multiple pustules which burst through several openings giving a cribriform appearance. Recurrent furunculosis or carbuncle suggests the patient has lowered general resistance, e.g. diabetes mellitus.

381. Carbuncle developing from a furuncle over the proximal phalanx of the little finger. This is also known as a whitlow.

Whitlow. An old term for a painful swelling on a finger, especially in the pulp. In more recent times it has applied to an infected swelling of the finger generally.

Carbuncle. Latin: *carbunculus*, a little live coal. In medicine the term carbuncle is applied to an inflamed condition involving the skin and producing a swollen, purple, red area of infection and inflammation.

Around the nail

382. Minor injury to the nail folds leads to cellulitis and, due to the relatively poor blood supply and poor capacity for distension of the semi-rigid tissues, the cellulitis often leads to abscess formation (376). This may be eponychial, paronychial or subonychial.

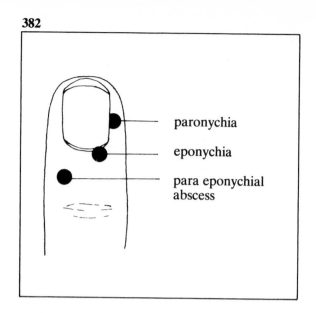

paronychia

eponychia

para eponychial abscess

383. Acute paronychia at the stage of cellulitis. The eponychium and paronychium show signs of inflammation.

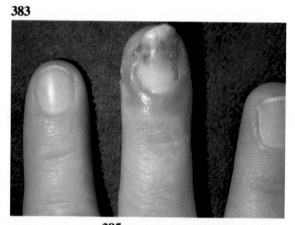

384 & 385. Acute paronychia at the stage of abscess. There is throbbing pain and tenderness over the abscess. The quantity of pus produced after lifting the nail fold is seen in **385**.

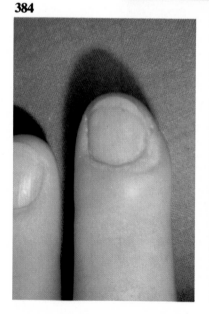

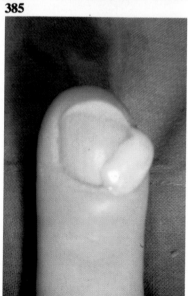

386. **Chronic paronychia in the left thumb.** Chronic inflammation, swelling of the nail fold and deformity of the nail itself. This occurs most commonly in those whose hands work in water, e.g. dish washers and barmaids.

Candida albicans or various bacteria may be cultured.

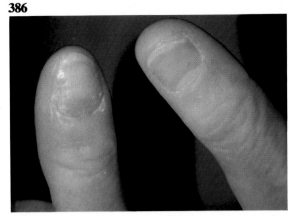

387. **Recurring paronychia in the left index finger of a 47-year-old woman with scleroderma.** As well as inflammation of the nail folds she had poor fingertip circulation, and thick, purpuric skin. Any patient with recurring or chronic paronychia should be suspected of having poor peripheral circulation or lowered general resistance.

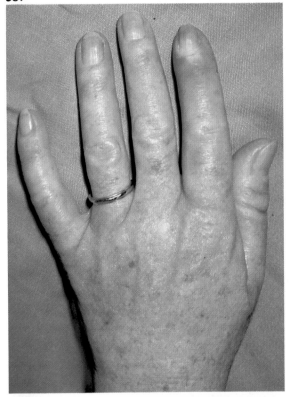

388. **A subungual abscess following penetration of a splinter beneath the nail (arrow).** Blood and pus are leaking from a hole the patient made in the nail to remove the splinter. The nail shows ridging from previous injury.

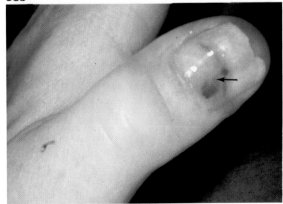

Pulp space

389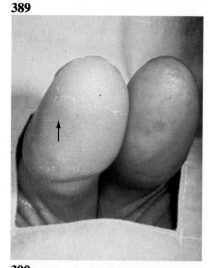

389. Distal pulp space infection of the right thumb (arrow). 'Felon', an early case. This patient presented with three days of increasing throbbing pain. Apart from local tenderness and slight swelling over the pulp there were no abnormal signs*. However, pus was subsequently drained from the pulp.

A small puncture wound from a prick produces cellulitis in the low resistance fat of the distal pulp. Suppuration occurs on account of the tension generated in the closed fascial space.

Note: the normal pulp is fluctuant.

390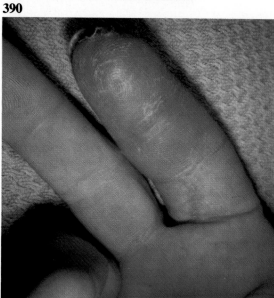

390. Severe pulp space infection of the right index finger. The entire distal pulp is tender and swollen, finger movements are restricted.

391. Dorsal swelling of the finger and hand.

392. Operative findings. Extensive necrosis of the pulp and involvement of the flexor profundus tendon and its synovial sheath.

391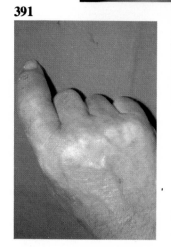

392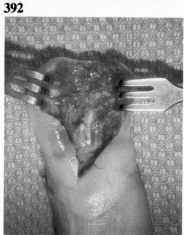

Felon. An obsolescent term for a severe form of whitlow or paronychia. *Felo* in Latin means a crime or wickedness.

393. Gangrene complicating a distal pulp space infection. Pain has been present for several weeks.

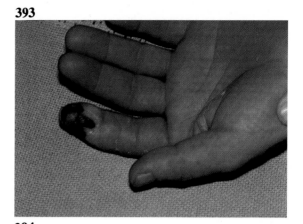

394. Lateral pulp infection. Persistent severe throbbing pain and localised tenderness indicated the presence of an abscess. This must be differentiated from paronychial infection which would be more dorsal.

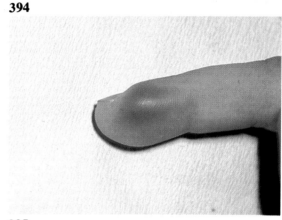

395. An apical infection. Severe pain is associated with a very small area of acute tenderness.

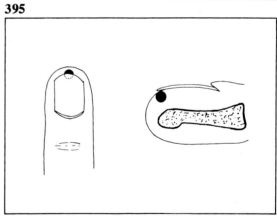

396. Recurrent apical pulp infection in a patient with Raynaud's disease. There is poor circulation at the fingertips.

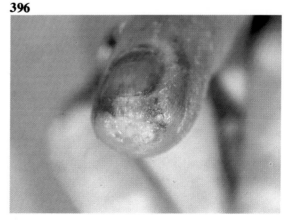

397

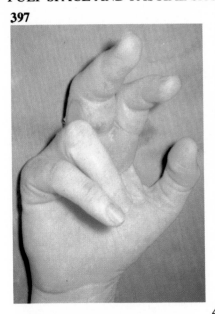

398

399

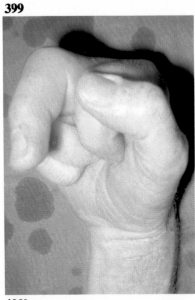

397–399. Proximal pulp space infection of the right middle finger. Pain, tenderness and swelling followed penetration of the pulp by a wooden splinter. The dorsal swelling is seen in **398** and the limitation of active flexion in **399**.

400a

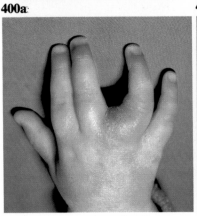

400b

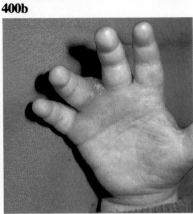

Fascial spaces

400a & 400b. Web space infection between the middle and ring fingers following a minor puncture wound. There was local tenderness and swelling in the web. The fingers are separated.

401a & 401b. Web space infection involving the flexor tendon sheaths of the ring and little fingers. Slight swelling in the web. Considerable dorsal swelling.

401a

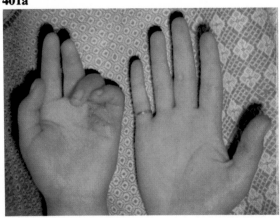

401b

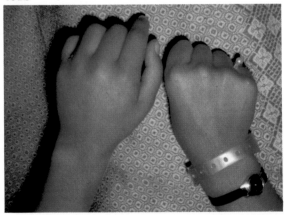

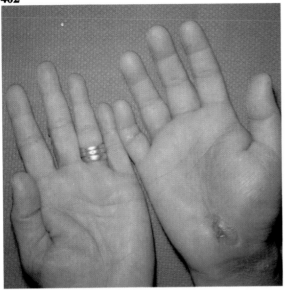

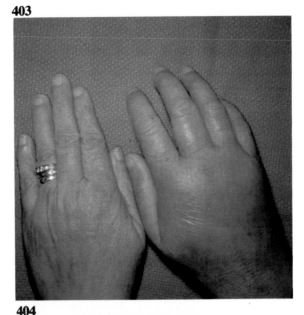

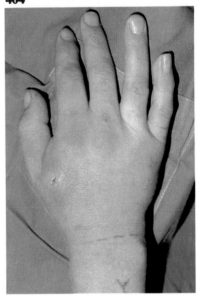

402. Thenar space infection. Four days after a puncture wound of the thenar crease there is pain, tenderness, swelling and restricted movement. The mid-palmar space was also involved.

403. Gross dorsal swelling.

404. A dorsal thenar web space infection.

   Mid-palmar space infection. See 'collar stud abscess', **380**.

405. Dorsal subcutaneous space infection following a bite over the metacarpo-phalangeal joint of the ring finger. There is extensive dorsal swelling.

406. A deep dorsal (subaponeurotic) space infection in an elderly diabetic. This abscess burst spontaneously and discharged foul smelling pus.

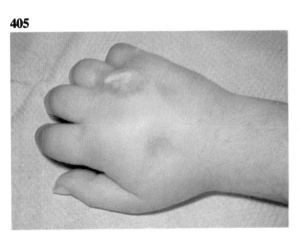

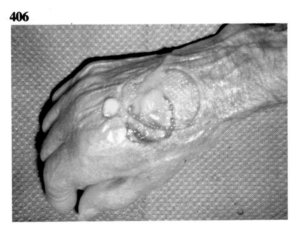

Tendon

407. Diagram of the flexor tendon sheaths. See **369**. Injury to the digital creases may result in infection of the tendon sheath.

Sites of palmar infections: (1) radial bursa, (2) thenar space infection, (3) flexor tendon sheath infection, (4) web space infection, (5) mid palmar space infection, (6) hypothenar space infection, (7) communication, (8) ulnar bursa; also the site of flexor tenosynovitis.

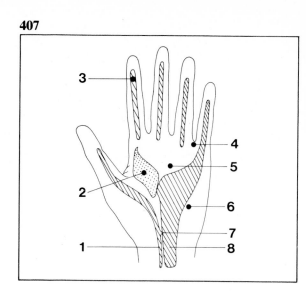

407

Flexor tenosynovitis. Kanavel described four signs. The finger is red and swollen, and held in slight flexion. There is tenderness along the sheath and severe pain on attempted extension.

Allen B. Kanavel (1874–1938). American general surgeon. Studied and wrote on infections of the hand in 1912. Kanavel's sign consists of tenderness over the base of the ring finger on the palmar side, occurring when an infection in the flexor sheath of the thumb has spread to the ulnar bursa. Kanavel stressed the importance of maintaining the hand in the position of function.

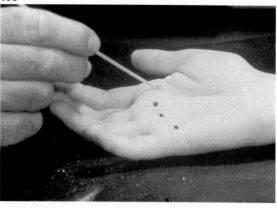

408

408. Testing for local tenderness over the proximal end of the flexor tendon sheath with a probe or swab stick.

409

409. Testing passive extension of the fingers. The hand rests on a table and gentle passive pressure is applied to the fingernail. In a patient with septic tenosynovitis such minimal movement of the flexor sheath produces exquisite pain.

410. Septic flexor tenosynovitis of the right middle finger of a veterinary surgeon who was bitten by a cat over the proximal interphalangeal joint crease.

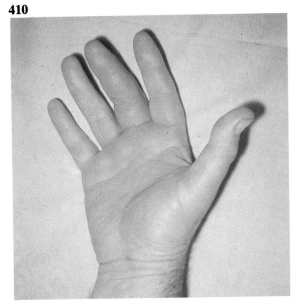

411. Flexion posture of the right middle finger and swelling of the finger and hand. The entire flexor tendon sheath was tender to palpation. Attempts to passively extend or flex the finger produced severe pain.

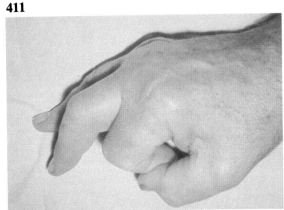

412. Septic tenosynovitis of the right thumb which has spread to involve the tendon sheaths of the other fingers. A thorn puncture wound of the right thumb pulp led to infection of the tendon sheath of the flexor pollicis longus. A week later this infection spread from the radial to the ulnar bursa and down to the fingers. Note the flexion posture of the thumb and fingers and swelling of the entire hand.

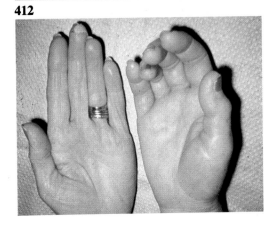

Bone

413. Acute osteomyelitis. Five weeks after penetration and infection of the lateral pulp space, the thumb pulp remained painful, tender and slightly swollen.

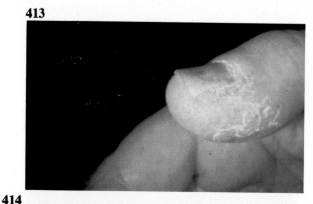

414. X-ray rarefaction of the distal phalanx.

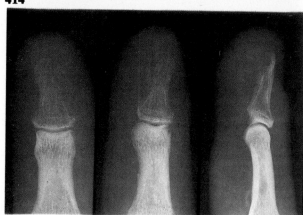

415. X-rays of the normal left thumb.

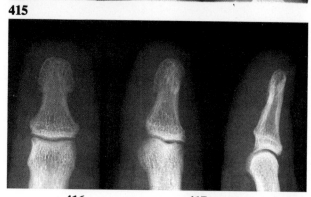

416 & 417. Acute osteomyelitis. This developed from pulp space infection. One week after a minor puncture wound from a splinter this patient, who lived by himself on an island, developed throbbing pain in the pulp. Over the next three weeks he treated himself by stabbing the pulp. On three different occasions he released pus and was relieved temporarily of pain. He presented six weeks later and was found to have osteomyelitis.

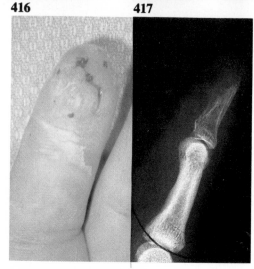

418. Chronic osteomyelitis with sequestrum in an Italian labourer was treated for acute osteomyelitis twelve months previously.

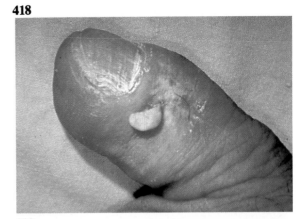

Joint

419. Septic arthritis of the distal interphalangeal joint of the right index finger. This man jammed his finger in a door and had the crushed wound sutured. Four weeks later he presented with persisting throbbing pain and purulent discharge from the distal interphalangeal joint.

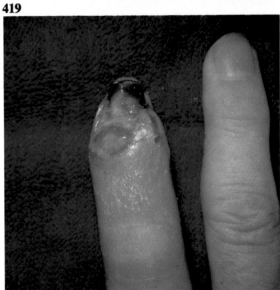

420. Osteomyelitis and septic arthritis with gross destruction of the distal interphalangeal joint. Early x-ray signs, i.e. at two to three weeks, include soft tissue swelling, rarefaction of juxta-articular bone and narrowing of the joint space from involvement of the articular cartilage.

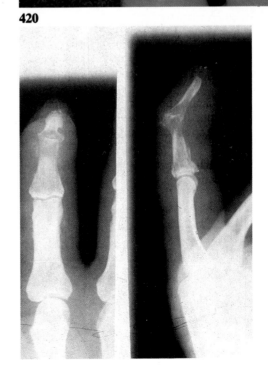

421

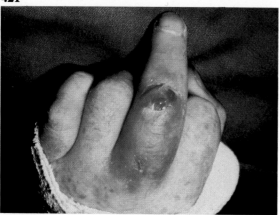

422

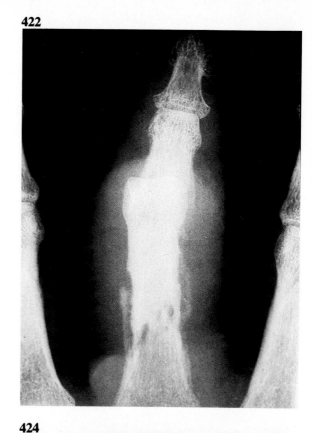

421. Septic arthritis occurring three weeks after a bite wound to the dorsal aspect of the proximal interphalangeal joint. The finger became increasingly painful until pus discharged. Bite wounds are often complicated by severe infection.

422. Joint destruction, dislocation and secondary osteomyelitis. X-rays.

423

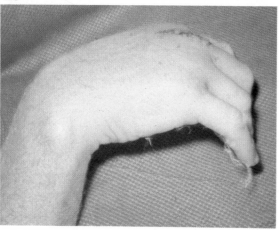

424

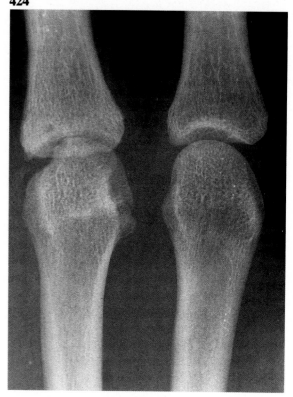

423. Septic arthritis following suture of a bite wound over the metacarpo-phalangeal joint of the left ring finger.

424. Narrowed joint space and irregular joint surface. X-ray.

Miscellaneous infections

Bacterial

425

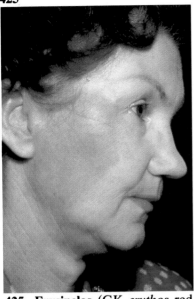

425. Erysipelas (GK *erythos* red, *pella* skin). A skin infection occurring most commonly on the face, hands and genitals. Severe constitutional reaction.

426

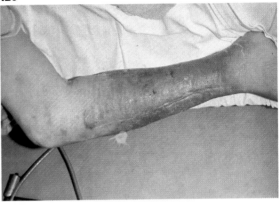

426. Erysipeloid ('fish handler's' disease) occurs almost exclusively in those handling fish and meat. After a scratch or cut, the area becomes inflamed and dark red or purple in colour. The infection lies in the skin or subcutaneous tissue. Constitutional symptoms are relatively slight.

427

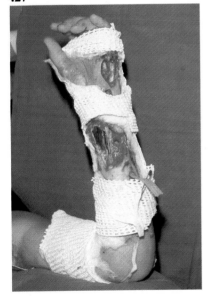

427. Gas gangrene in a 26-year-old labourer. Suture of a solid contaminated wound was followed by application of too tight a plaster. The combination of contamination and ischaemia resulted in gangrene. Amputation was performed.

428

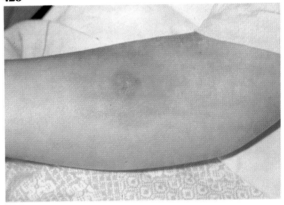

428. Tuberculosis. Papulonecrotic tuberculid, a tuberculous papule with necrosis.

Gas gangrene can be fatal.

Radiologically, gas may show in the muscle bundles. Gas can also show in the absence of infection, being introduced by way of surgery.

429. Tuberculosis presenting as a persistent sinus in the thenar crease. This Yugoslav meat slaughterman contracted bovine tuberculosis following penetration with a carcass bone. Tuberculous tenosynovitis was found at operation. The ring finger shows scarring from previous injury.

429

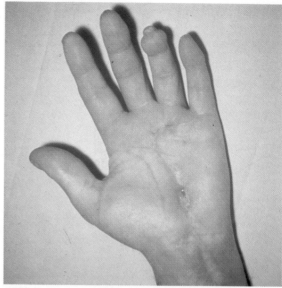

Leprosy. Whether of the lepromatous or tuberculoid type, leprosy presents predominantly as nerve or skin lesions.

430. Gross enlargement of the right median nerve at the wrist. The size of the nerve is indicated by the forceps. A similar enlargement can occur in the ulnar nerve at the olecranon.

430

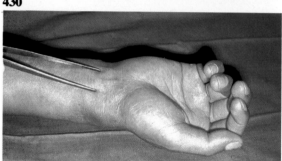

431. Leprosy ulcer over the left wrist. This is not a trophic ulcer but arises in a raised red leprosy plaque.

431

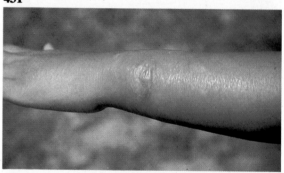

432. Anthrax. Though known as malignant pustule, anthrax is rarely fatal. A gangrenous carbuncular lesion in the forearm of a cattle worker.

432

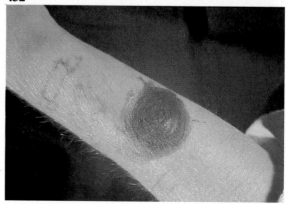

433

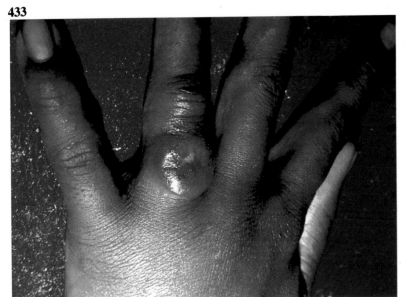

434

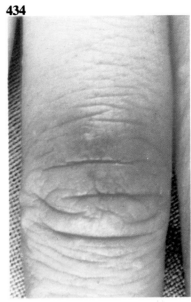

433. Granuloma annulare. A chronic non-infectious granuloma which presents as a reddish firm nodule on the hands or feet. It can ulcerate. Pathologically there is a focal degeneration of collagen with reactive inflammation and fibrosis.

Differential diagnosis. Rheumatic nodule, necrobiosis lipoidica.

434. Gonorrhoea. Vesicopustule of the finger. After instrumentations or manipulation of infected tissue, bacteraemia may occur and cause skin lesions, septic arthritis or tenosynovitis.

435

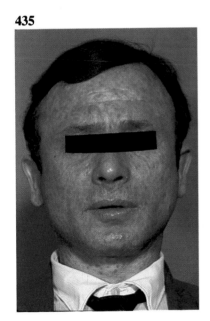

436

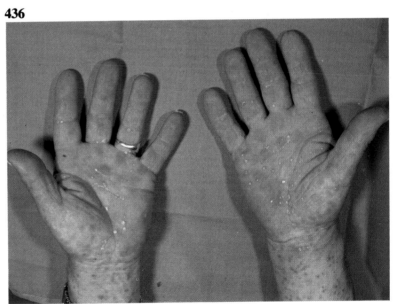

435 & 436. Syphilis of the hands and face. The skin lesions are so variable that they imitate many disorders. The first skin signs are reddish macules on the palms, soles and face. Later these may change to coppery papules. They are never vesicular or bullous and they do not itch.

Viral

437

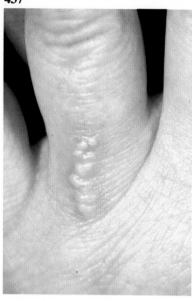

438

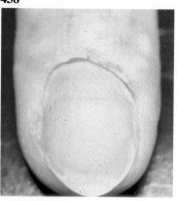

438. Herpes simplex of the nail fold. The vesicles take 24 to 48 hours to appear – before then, herpes may be mistaken for a distal pulp infection.

Warts. These are virus induced pseudo tumours. See **455**.

437. Herpes simplex of the dorsal skin of the finger.

Fungal

439

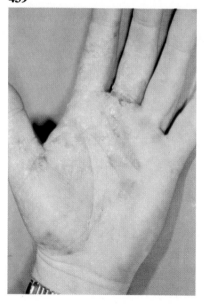

440

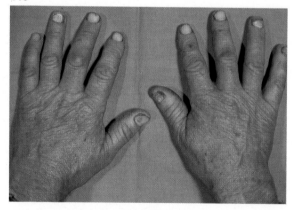

440. Tinea unguium. Irregularity of the nails, particularly of the right thumb and little finger. This fungal infection begins at the distal end of the nail plate and spreads proximally. The under surface of the nail may be seen to be hyperkeratotic.

The colour of the nail does not alter unless bacterial invasion takes place; it then becomes greenish. Nevertheless there is a loss of normal nail lustre.

439. Tinea manuum. Ring worm of the hand, which can present as scaling or as vesiculopustular eruption. Both forms are seen in the hand, and the feet are also involved.

441. Fungal infection of glans penis and nails.

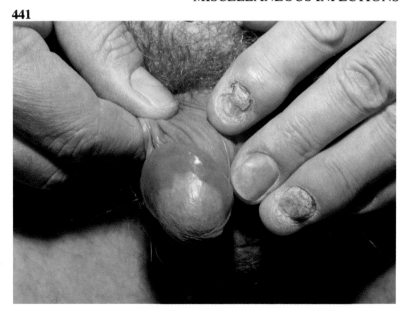

Parasitical

442. Scabies. Characterised by intra-epidermal burrows, follicular papules and severe itching which is worse at night. Occurs in the finger webs and wrist creases. Highly contagious.

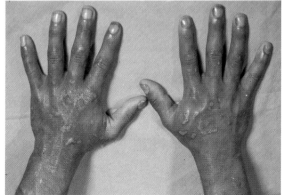

443. Pediculosis (lice infestation). Scratch marks and secondary impetigo may be the only signs.

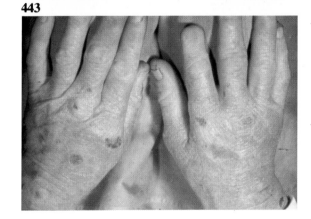

Differential diagnosis of infection

444. Allergic reaction to a rose thorn. This resembles a streptococcal blister. The adjacent finger shows an old subungual haematoma.

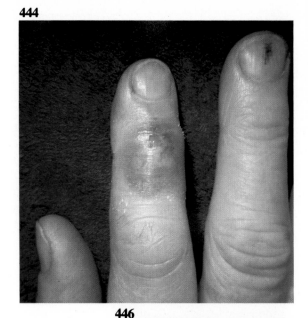

445 & 446. The signs of acute soft tissue inflammation without pyrexia. The white spots indicate gout. The serum uric acid level was increased. Gout may also mimic acute septic arthritis.

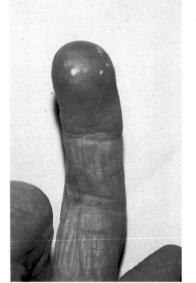

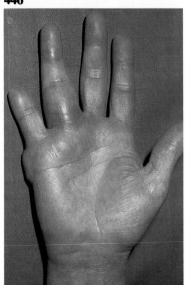

447. Squamous cell carcinoma of the nail bed. This was misdiagnosed for nine months as fungus infection. The diagnosis was obvious when the nail was removed. By this time there were metastatic lymph nodes in the axilla.

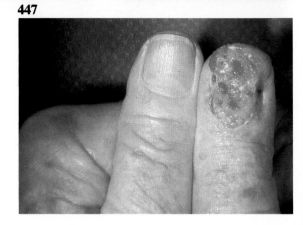

448

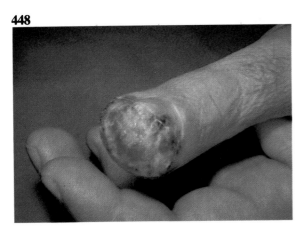

449

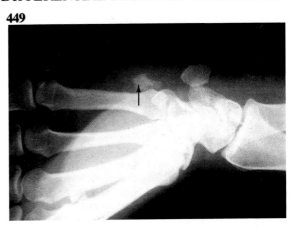

448. Amelanotic melanoma. This presented as poor healing two months after a crush injury. The diagnosis was eventually made by biopsy.

449. Acute calcification. Though rare, acute calcification can occur in wrist tendons and present as soft tissue inflammation.

450a, b. Secondary carcinoma from primary carcinoma of the lung. This patient presented with an inflamed swollen fingertip. The distal phalanx is all but destroyed.

450a

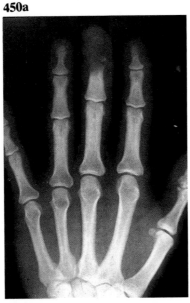

450b

451. Self-inflicted sores. This patient maintained his sores by repeated use of a toothpick.

452. This patient was found to be injecting his hand with barbiturates.

451

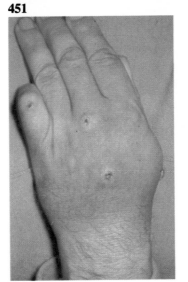

452

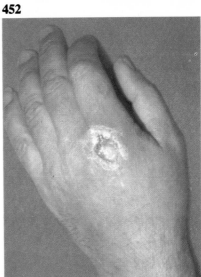

143

5. Tumours

Hand tumours may be classified as pseudo, benign or malignant tumours.

Most tumours of the hand are not true neoplasms but the result of classic disease states, e.g. trauma, infection, inflammatory disease, metabolic disorder, degenerative disorder or congenital anomaly.

Tumours of the hand

TISSUE	PSEUDO-TUMOURS	BENIGN TUMOURS	MALIGNANT TUMOURS
Skin	Wart (verrucae vulgaris) 455 Sebaceous cyst 456 Hypertrophic scar (keloid) 457 Epidermoid inclusion cyst 458–462 Pyogenic granuloma 463 Rheumatoid nodule 464	Keratosis (seborrhoeic, solar, senile) 513 Dermatofibroma 514 Kerato-acanthoma 515–516	Bowen's disease 559 Basal cell carcinoma 560 Squamous cell carcinoma 561–563 Melanoma 564–570
Subcutaneous tissue, fat	Gouty tophus 465–466 Haematoma, seroma 467 Foreign body granuloma 468–469	Lipoma 517–520	Liposarcoma 579–580
Fascia	Dupuytren's nodule 470 Knuckle pads 471	Fibroma 521	Fibrosarcoma Epithelioid sarcoma
Tendon	Tendon stump 472 Rheumatoid nodule 473 Tendon cyst 474 Synovitis 475–478	Giant cell tumour 522–526 Xanthoma 527	Malignant giant cell tumour of tendon sheath
Muscle	Congenital anomaly Palmaris longus 479–480 Extensor digitorum brevis 481–482 Hypertrophy of lumbrical muscle 483 Haematoma	Leiomyoma 528 Rhabdomyoma	Leiomyosarcoma Rhabdomyosarcoma

Tumour. *Tumor* in Latin means a swelling. More recently the tendency has been to restrict the use of tumour to a new growth.

TISSUE	PSEUDO-TUMOURS	BENIGN TUMOURS	MALIGNANT TUMOURS
Blood vessel	Aneurysm **484–485**	Haemangioma (capillary cavernus) **529–532**	Haemangiosarcoma **581**
	Lymph cyst **486–487**	Haemangiolymphoma **533–534**	
	Haemangioma **488**		
	Aberrant radial artery **489**	Glomus tumour **535–536**	
	Varix, A-V fistula		
Nerve	Neuroma **490**	Neurilemmoma (Schwannoma) **537**	Malignant neurilemmoma
		Neurofibroma **538**	
		Neurofibromatosis (Von Recklinghausen's disease) **539**	
		Neurofibrolipoma **540**	
		Haemangioma **541**	
Joint	Ganglion **491–501**	Giant cell tumour of joint synovium	Synovial sarcoma
	Mucous cyst **502–504**		
	Knuckle bursa **505**		
Bone	Heberden's node **506**	Enchondroma **542–543**	Osteogenic sarcoma **571–575**
	Exostosis and callus **507–508**	Osteochondroma **545–546**	Chondrosarcoma **576**
	Carpo-metacarpal boss **509–510**	Giant cell tumour **547–552**	Ewing's tumour **577**
	Multiple hereditary exostosis	Aneurysmal bone cyst **553–555**	Secondary carcinoma **578**
	Dislocated joint	Osteoid osteoma **556**	
		Periosteal chondroma **557**	
		Bone cyst **558**	
Miscellaneous	Paraffinoma **511**		
	Myxoma **512**		

Diagnosis

Presentation of hand tumours

Tumours of the hand present in one of five ways:
(1) Lump
(2) Fracture
(3) Pain
(4) Nerve palsy
(5) Incidental x-ray finding

Method of diagnosis

(1) History

When was it first noticed? A ganglion, pyogenic granuloma, implantation cyst may follow injury. A haemangioma may be present from birth.

Is it painful? e.g. glomus tumour, osteoma, neurofibroma etc.

Is there associated systemic disturbance? e.g. gout, rheumatoid arthritis.

Has there been rapid growth? e.g. malignancy.

Did the lump arise suddenly after injury? e.g. haematoma, aneurysm, ganglion, Dupuytren's nodule.

Is there variation in size? e.g. ganglion, mucous cyst.

(2) Examination

Is it cystic, solid or fluctuant?

Is it translucent? e.g. ganglion, mucous cyst, lipoma.

Is it large? e.g. lipoma.

Is it small? e.g. glomus tumour.

Are there multiple tumours? xanthoma, neurofibromatosis.

Is it bony? see x-ray features.

Is there involvement of the regional lymph nodes? e.g. squamous cell carcinoma.

Aim of diagnosis

(1) Which tissue is involved? Soft tissue – skin, subcutaneous tissue etc. or skeleton – bone, joint.

Some soft tissue tumours involve bone and present distinct radiological findings.

(2) What is the pathological process?
   Pseudo tumour – ganglion.
   Benign tumour – giant cell tumour.
   Malignant tumour – secondary carcinoma.

(3) X-ray. This may show:

(a) bone tumour, primary or secondary, with or without soft tissue swelling.

(b) soft tissue tumour with secondary cystic change in bone, e.g. epidermoid cyst, giant cell tumour, glomus tumour.

(c) calcification in the soft tissues, e.g. haemangioma, lipoma, calcinosis circumscripta, juvenile aponeurotic fibroma.

Diagnostic features of bone tumours

(a) Site in the hand

Distal phalanx – enchondroma, secondary cystic change from epidermoid cyst, glomus tumour, etc.

Proximal phalanx – enchondroma, ecchondroma.

Carpus. It is rare to have an enchondroma or ecchondroma in the carpus.

(b) Site in the bone

Epiphysis – giant cell tumour.
Metaphysis – ecchondroma.
Diaphysis – enchondroma.

(c) Radiological features of bone cysts

Cyst + soft tissue tumour – xanthoma.
Stippling – enchondroma.
Trabeculation – aneurysmal bone cyst.
Small – osteoid osteoma.
Large – aneurysmal bone cyst.
Bone destruction – sarcoma.
Bone sclerosis – osteoid osteoma
                – infection (osteomyelitis)
Clear cyst + good cortex – epidermoid cyst.
Irregular calcification – chondroma.

(4) Biopsy. The nature of most hand tumours is usually obvious. Where there is any doubt, an histological examination should be made. See **447, 448**.

Many tumours have a predilection for certain areas of the hand.

Diagnose by history, examination, investigation and biopsy. First determine the tissue of origin and then the most likely pathological state.

453

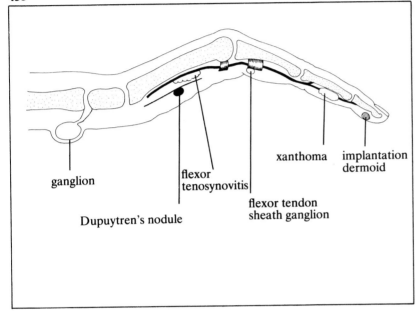

453. Palmar hand tumours.

ganglion

flexor tenosynovitis

Dupuytren's nodule

flexor tendon sheath ganglion

xanthoma

implantation dermoid

454

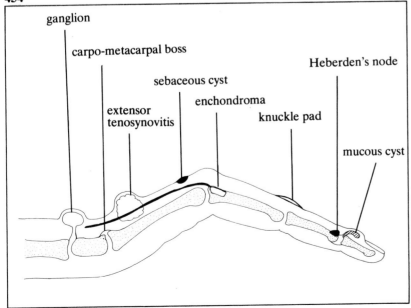

454. Dorsal hand tumours.

ganglion

carpo-metacarpal boss

sebaceous cyst

enchondroma

extensor tenosynovitis

knuckle pad

Heberden's node

mucous cyst

Common hand tumours

Skin
Epidermoid cyst; keratoses; basal cell carcinoma; squamous cell carcinoma.

Subcutaneous
Ganglion; xanthoma.

Skeletal
Enchondroma; osteochondroma.

Non-neoplastic disorders which cause tumours

Trauma
Haematoma; seroma; ganglion; epidermoid inclusion cyst; foreign body; neuroma; tendon stump; hypertrophic skin scar; bone stump or exostosis; knuckle bursa or aneurysm.

Infection
Warts (verruca vulgaris); abscess; pyogenic granuloma.

Chronic inflammatory states and collagen disorders
Rheumatoid disease causing synovitis as well as skin, subcutaneous and tendon nodules; non-specific villonodular tenosynovitis; Dupuytren's fasciitis; keratoacanthoma, or Boeck's sarcoid.

Metabolic disorders
Gouty tophus; xanthoma.

Degenerative disorders
Hypertrophic osteo-arthropathy with Heberden's nodes; exostosis, ganglion; senile and seborrhoeic keratosis.

Congenital anomaly
Haemangioma; ganglion; anomalous bone or muscle formation or arteriovenous fistula.

Pseudo tumours

Pseudo tumours are swellings – they are not true neoplasms.

Skin and fat

455

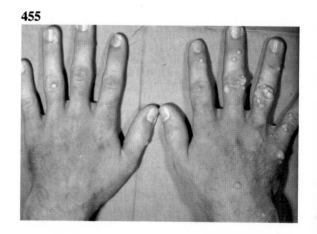

455. Warts (verrucae vulgaris). More than 50% of these virus induced pseudo tumours occur in the hand, especially in children. The common raised warts show hyperkeratosis, acanthosis and papillomatosis. (Acanthosis = hypertrophy of the prickle cell layer).

456

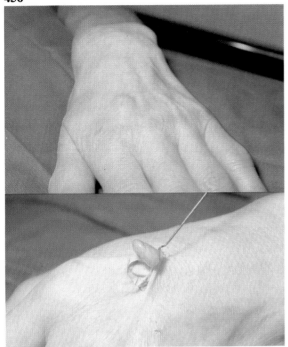

456. Sebaceous cyst presenting as a dorsal hand tumour.

457

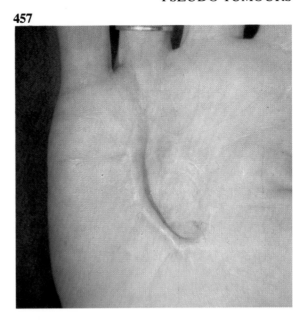

457. Hypertrophic scar – following laceration of the palm in a 9-year-old school girl. Scars are prone to hypertrophy when they lie perpendicular to a flexion crease, especially in the young growing hand after burns.

458

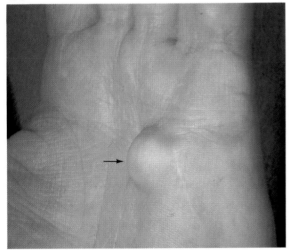

458. Epidermoid inclusion (implantation) cyst. See arrow. Presenting as a hard, rounded, semi-fluid, elastic, subcutaneous tissue tumour. The lump over the base of the little finger is a Dupuytren's nodule. It is attached to a thickened band of fascia.

459

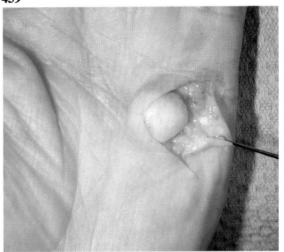

459. Operative findings. The cyst consists of a cavity filled with debris and cholesterol. It is surrounded by a wall of epithelium resembling skin.

Differential diagnosis. Foreign body granuloma, ganglion, xanthoma, Dupuytren's nodule.

460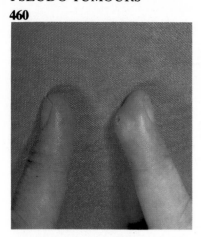

461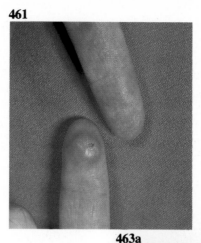

462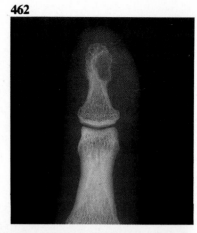

460–462. Implantation dermoid cyst. Presenting as a slow growing tumour; a splinter had been removed ten years before.

Epidermoid, implantation, inclusion or post-traumatic cysts. These are probably caused by traumatic implantation of epidermal tissue which then grows into a cyst. They occur months or years after penetrating or crush injury, on the palmar surface, usually the fingertip, and occasionally in amputation stumps. They are seen in those whose fingers are prone to trauma, such as tailors and machinists. The cysts vary in size from 1–20mm. They contain keratinous debris and are lined with squamous epithelium. On x-ray the cysts present as lucent defects in the tufts of the distal phalanx. They are usually intra-osseous or seem to be secondary erosions from a subcutaneous cyst.

Differential diagnosis: glomus tumour, enchondroma, metastasis.

463a, b. Pyogenic granuloma. A mound of granulation tissue initiated by a puncture wound of the skin.

Subcutaneous tissues

Rheumatoid nodules. May involve the skin, fascia, bursae or tendons. 20% of patients with rheumatoid arthritis have such nodules. The nodules may antedate the appearance of overt rheumatoid disease.

464. Rheumatoid nodule presenting as lobulated firm swellings of the upper forearm and elbow.

463a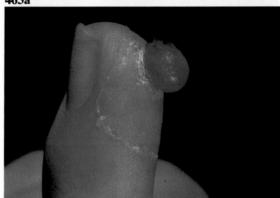

463b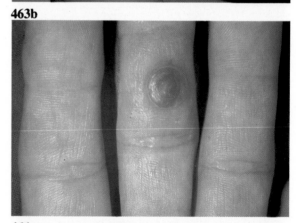

464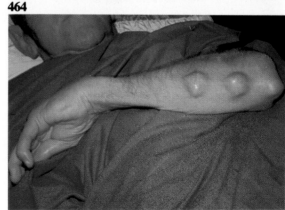

465

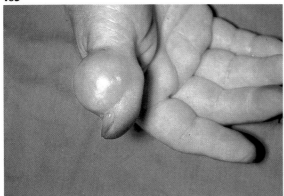

465. Gouty tophus, ulcerating through the skin. The white discharge contains a uric acid crystal. Gouty, chalky, deposits can occur in any tissue.

466. Gout. There are irregular erosions around the neck of the middle phalanx and the base of the distal phalanx. There is considerable soft tissue swelling. The articular surfaces are normal.

Haematoma or seroma. A haematoma can occur rapidly after injury but if the skin is thick and contains considerable subcutaneous fat, bruising may not appear for several days. A seroma follows in two or three weeks if the haematoma does not reabsorb.

467. Seroma. Developing three weeks after a blow on the dorsum of the hand.

466

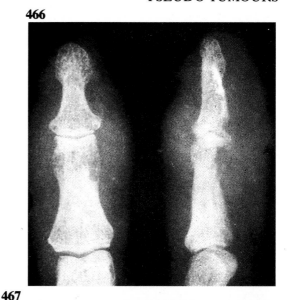

467

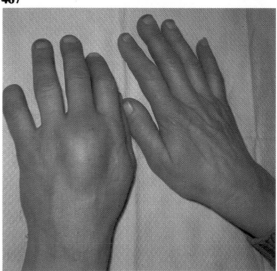

468. Foreign body granuloma presenting as tender subcutaneous lump several weeks after penetration of a wooden splinter.

469. Operative findings. Splinter surrounded by inflammatory granulation tissue.

468

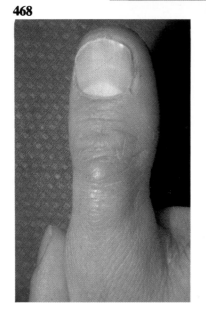

469

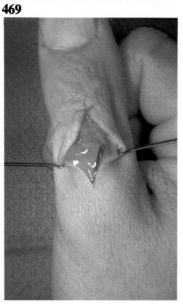

Fascia

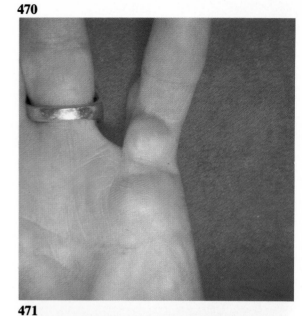

470. Dupuytren's nodules associated with a fascial band. Such nodules occur most commonly in the distal palm over the bases of the ring and little fingers. They present as firm to hard subcutaneous nodules which can be tender if the process is acute. The fascial bands can cause joint contracture.

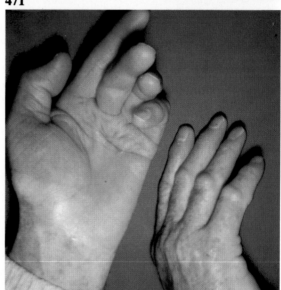

471. Knuckle pads. Non-inflammatory hyperplasia of fibrous tissue, akin to Dupuytren's fasciitis, occurs over the dorsal aspect of the proximal interphalangeal joints in patients with a Dupuytren's diathesis. Knuckle pads can also occur over the dorsal aspect of the other finger and thumb joints. They are also seen in shearers. (See Chapter 9).

Tendon

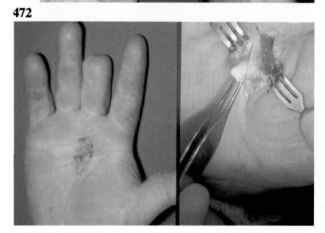

472. Retracted flexor tendon stumps presenting as a tender palpable lump in the palm. The flexor tendons were divided during amputation of the middle finger. Such retracted stumps are usually surrounded by a pouch of fluid.

473

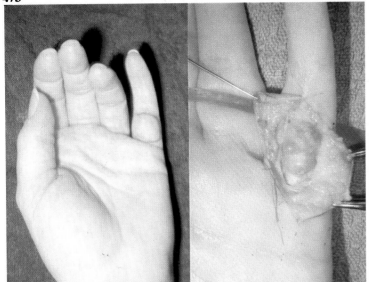

474

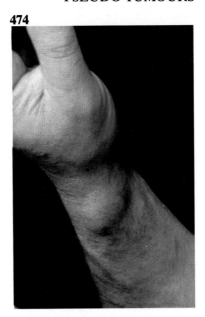

473. Rheumatoid nodule in flexor tendon. The rheumatoid degenerative process may involve the tendon or the tendon sheath. There is restricted tendon glide when the nodule catches on the tendon pulley, and the tendon may eventually rupture from attrition.

474. de Quervain's syndrome presenting as a painful tumour.

Fritz de Quervain (1868–1940). Swiss general surgeon who described his syndrome, i.e. occupational chronic tenovaginitis, in 1895 whilst working in the watch-making district of La Chaux-de-Fonds.

475

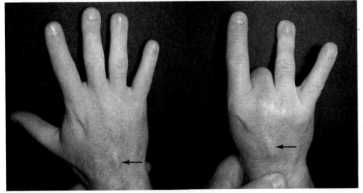

476

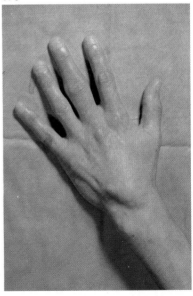

475. Tendon cyst. Cysts or nodules may form on flexor or extensor tendons. They are small, hard, and move with the tendon.

Synovitis, villo nodular tenosynovitis. Pseudo tumours of proliferative synovitis may arise from a tendon sheath or joint in non-specific, rheumatoid or tuberculous conditions. They usually present as tumours of the extensor or flexor tendons.

476. Extensor tenosynovitis. Probably rheumatoid, asymptomatic. Examining fingers cannot get above the swelling, which extends beneath the extensor retinaculum.

Differential diagnosis. Ganglion. See also **491–501**.

477

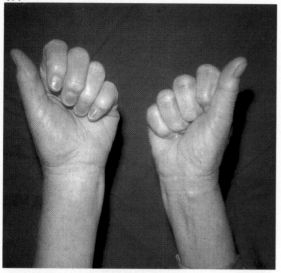

478

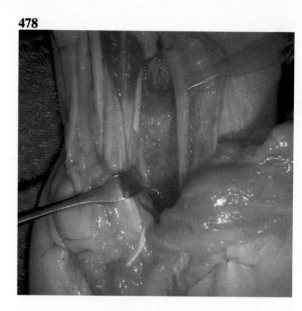

477. Flexor tenosynovitis of the left wrist. Pain and swelling above the wrist. Secondary carpal tunnel compression of the median nerve.

478. Operative findings. The tenosynovium has been stripped from the flexor tendons.

Muscle

Congenital anomalies of muscle may present as a tumour.

479

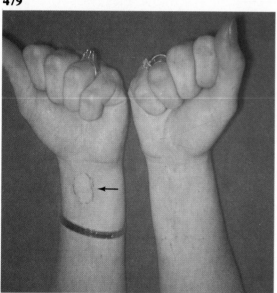

480

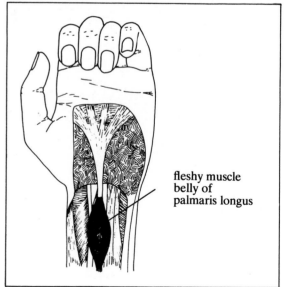

fleshy muscle belly of palmaris longus

479. Palmaris longus muscle (arrow) presenting as a palpable swelling proximal to the wrist.

480. Diagram.

481

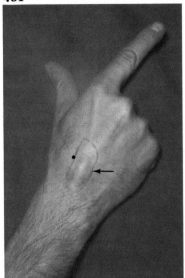

482

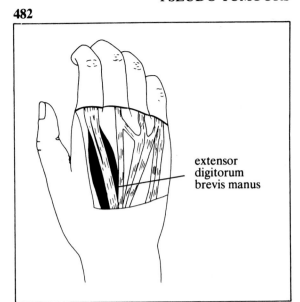

extensor
digitorum
brevis manus

481. Extensor digitorum brevis manus. A congenital anomaly, usually of extensor indicis and bilateral in one-third of cases. Presents as a soft, fusiform swelling which moves in the line of the extensor tendon. Occasionally presents as discomfort after heavy work. *Differential diagnosis:* Ganglion, soft tissue tumour.

482. Diagram.

483

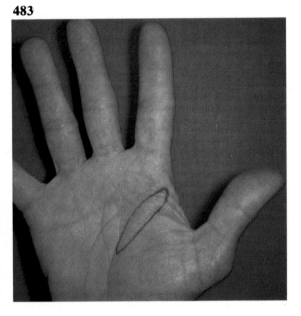

483. Hypertrophy of the first lumbrical muscle presenting as a palmar tumour.

There may be pain from pressure on the digital nerve to the radial side of the index finger. This nerve lies on the lumbrical muscle in a very superficial and vulnerable position.

A muscle haematoma can follow trauma and present as a tumour.

Blood vessels

484

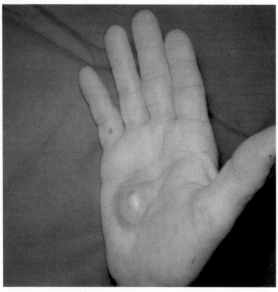

485

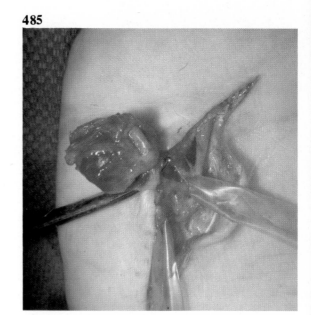

484. Traumatic aneurysm presenting as a blue, firm, non-pulsatile, subcutaneous hand tumour. A glass puncture wound of the palm was followed immediately by swelling. This became pulsatile, but by the time of operation three months later this sign had disappeared, from thrombosis.

485. Operative findings. Traumatic aneurysm of superficial palmar arch.

486

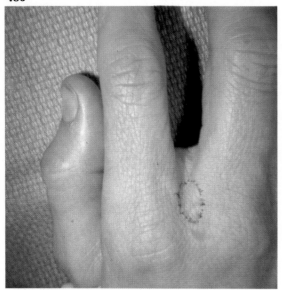

487

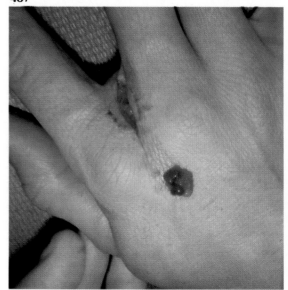

486. Lymph cyst in an 18-year-old typist. A small, palpable, firm, mobile, subcutaneous lump.

487. Lymph cyst. Injury prior to its removal has filled it with blood.

488

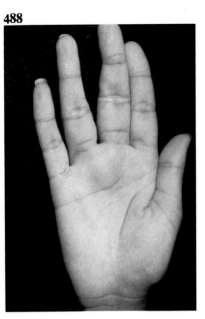

489

488. Haemangioma presenting as a bluish soft compressible swelling.

Arteriovenous fistula. See **687**.

Varix. Varicose veins of the digit. See **686**.

489. Abnormal course of the radial artery presenting as a pulsatile tumour at the wrist.

Nerves

Neuromas are of two types: (i) amputation neuroma or (ii) neuroma in continuity. Their characteristic clinical features are a 'pins-and-needles' or an electric shock sensation when they are knocked. They may also present as a lump.

490. Neuroma. Complete division of median nerve, with scar bridging the gap between the two ends of the nerve.

490

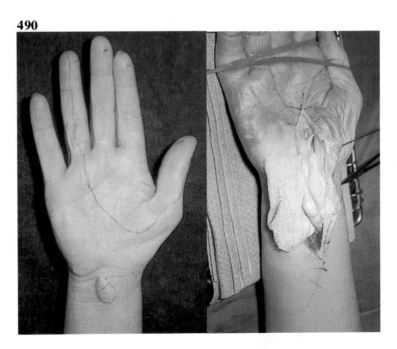

Joints

Ganglion. A ganglion is a synovial cyst arising from a joint or tendon sheath. It is the most common tumour of the hand. There are three main sites:

(a) the wrist, on either the dorsal or volar aspects,

(b) an interphalangeal joint, particularly in association with osteo-arthritis (see mucous cyst, **502**),

(c) the flexor tendon sheath usually at the level of the metacarpo-phalangeal joint.

The cyst is soft, fluctuant, translucent and often multilocular. It can vary in size, being quite small or very large. Aspirated fluid is clear, colourless and jellylike.

491. A dorsal wrist ganglion. Examining fingers can delineate the proximal limit of the swelling. See **476**.

492. Soft tissue swelling. X-ray.

493. Operative finding. A ganglion arises from the wrist joint and presents to the surface between the extensor tendons.

494. A volar wrist ganglion. These can present proximal to the wrist but they can usually be tracked down the flexor carpi radialis tendon to the wrist joint itself.

Differential diagnosis. Osteo-arthritis of the first carpo-metacarpo joint (see **839**) or synovitis of the flexor carpi radialis tendon.

Paul of Aegina (625–690). Byzantine surgeon and obstetrician who wrote probably the first account of a ganglion.

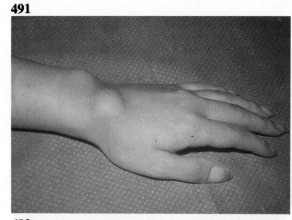

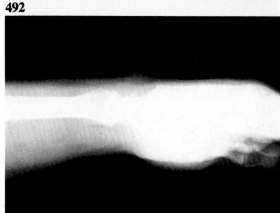

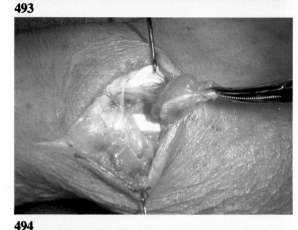

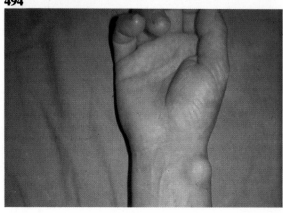

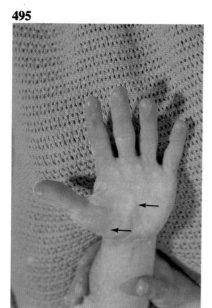

495. A wrist ganglion presenting as thenar palsy in a 35-year-old surgeon. Arrows point to the ganglion and the thenar muscle wasting.

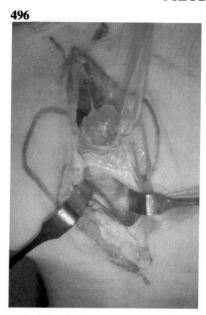

496. Operative findings. A ganglion compressing the thenar branch of the median nerve.

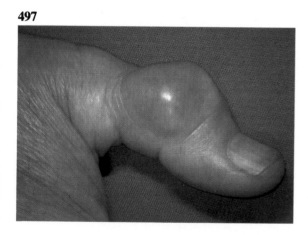

497. An interphalangeal joint ganglion in a 70-year-old man. Mucous cyst, see **502**.

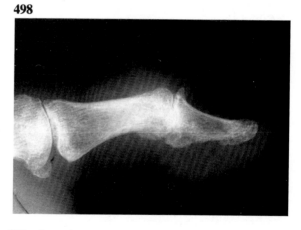

498. Associated osteo-arthritis. X-ray.

Ganglion. *Ganglion* in Greek means a knot or a tumour under the skin. The ancient Greeks did not distinguish between tendon and nerve. Ganglia thus could arise either from tendon or nerve.

499. A flexor tendon sheath ganglion. Also called a volar retinacular ganglion because it arises from the fibrous retinacular pulley near the metacarpophalangeal joint. It presents as a tender, hard, fixed, palpable nodule.

Differential diagnosis. Sesamoid bone.

499

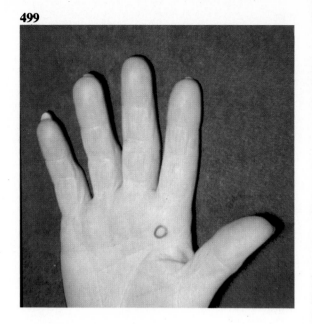

500

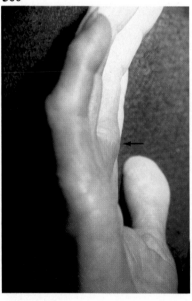

500. Flexor tendon sheath ganglion. Lateral view. Arrow.

501

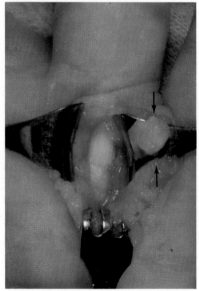

501. Operative findings. Ganglion marked by two arrows.

502. Mucous cyst. This small firm ganglion-like cyst arises from the dorsal aspect of the distal interphalangeal joint usually in the middle finger of adult females. Osteo-arthritis of the joint is always present. It may produce grooving of the nail or a thick discharge from under the eponychium.

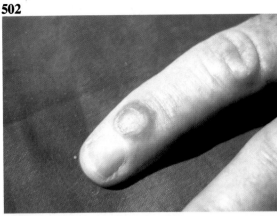

503 & 504. Mucous cyst presenting as grooving of the nail and subeponychial discharge.

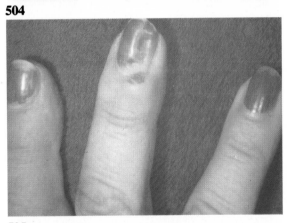

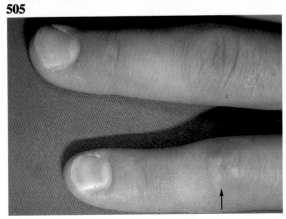

505. Knuckle bursa. A cystic swelling over the dorsal aspect of the proximal interphalangeal joint occurring after injury.

Bone

506. Heberden's node of the distal interphalangeal joints. For the x-ray see **837**.

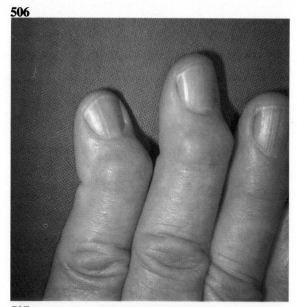

507. A bony hard tumour of the right index finger.

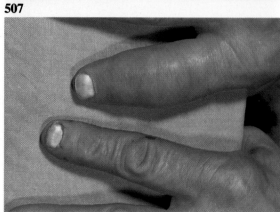

508. Callus around a fracture site. Multiple hereditary exostoses, Madelung's deformity, and a dislocated joint can also present as a hand tumour.

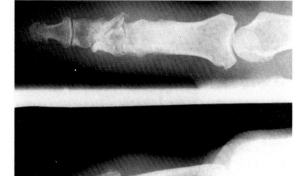

William Heberden (1710–1801). English physician. A Greek and Hebrew scholar. He wrote his case notes in Latin – these, his 'commentaries', were published after his death in 1802 and translated into English by his son. *Heberden's nodes.* Hard nodules adjacent to the distal interphalangeal joints of the fingers, occurring in osteo-arthritis.

Otto Wilhelm Madelung (1846–1926). German professor of surgery. In 1878 he described the deformity of the wrist occurring in young women, probably due to a defect in the development of the distal radial epiphysis.

509

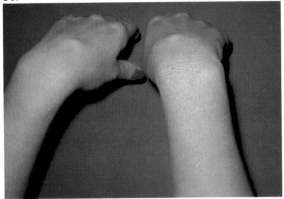

510

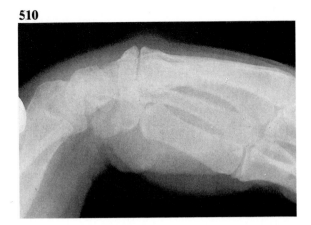

509 & 510. Carpo-metacarpal boss or separate styloid process of the third metacarpal, presenting as a bony hard tumour at the wrist. Misdiagnosed as a ganglion, often because of an overlying bursa.

Miscellaneous

511

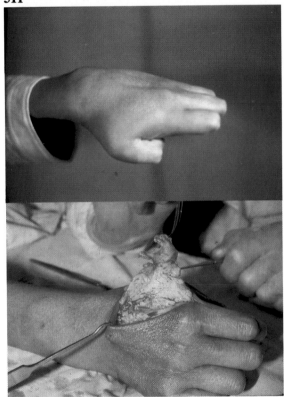

511. Paraffinoma. Paraffin was injected to fill the wasted first web space in a Chinese leper who had ulnar nerve palsy.

512. Myxoma. 65-year-old male. Four year history of swelling of the left thumb pulp. Slow growth.

512

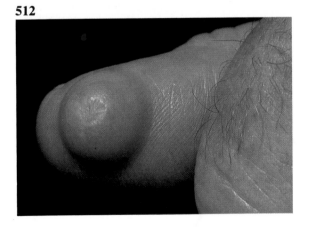

Benign tumours

Skin

513. Keratosis, solar keratosis, senile keratosis. Small, firm and scaly plaques are found on the extensor aspect of the hand in the older age group. They are found also on the face and other areas of the body exposed to sun and weather. These lesions are pre-malignant.

Differential diagnosis. Squamous cell carcinoma.

514. Dermatofibroma. Subepidermal nodular fibrosis (arrow). This is a fibrous proliferation in response to minor trauma. It is reddish or yellowish in colour and presents as a raised, firm nodule generally less than 1cm in diameter.

515 & 516. Kerato-acanthoma. Presents as an umbilicated skin nodule on the dorsum of the hand. It varies in size from 0.5–2cm in diameter, reaching its maximal size at 4–6 weeks. It may regress without treatment and is hence called a 'self healing cancer'.

Differential diagnosis. Squamous cell carcinoma.

513

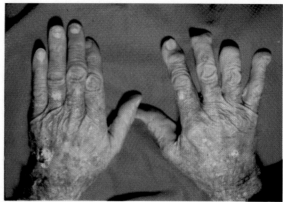

514

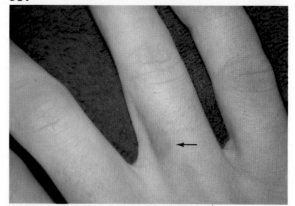

515

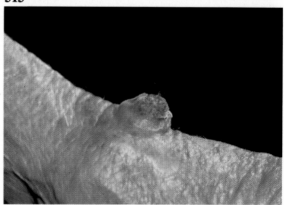

516

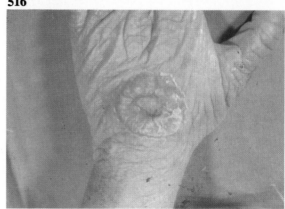

Subcutaneous tissue, fat

Lipomas are soft, flabby, lobulated, encapsulated, translucent tumours with pseudo fluctuation. They lie either in the subcutaneous or subfascial plane, usually on the volar aspect.

517. Lipoma of the mid-palm presenting as an ill defined translucent swelling (arrow).

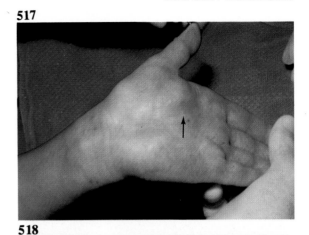

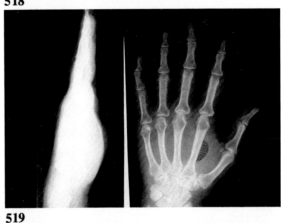

518. Translucent swelling. X-ray.

519. Operative finding. An extensive subfascial lipoma.

Differential diagnosis. Xanthoma, haemangioma.

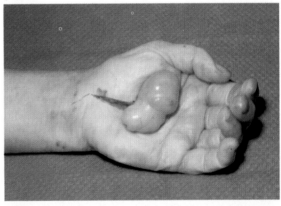

520. Lipoma of the left index finger and palm.

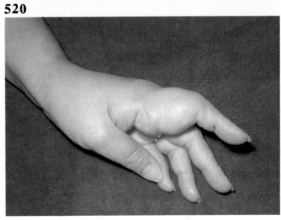

Fascia

Fibromas are rare tumours of the hand but they may arise from the skin, subcutaneous tissue or the fascia on either volar or extensor aspects.

521. **Operative finding** of a fibroma presenting as a firm, palpable lump in the palm.

521

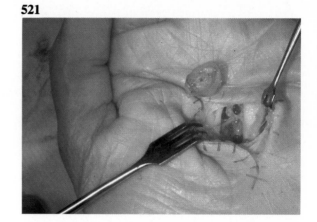

Tendon and tendon sheath

Giant cell tumour. This is a benign, encapsulated tumour of unknown aetiology, resembling a xanthoma. It arises from the white tissues of the hand, most commonly the tendon sheath, joint ligament or joint synovium. It is firm, fixed to the deep fibrous tissue of origin, and is differentiated from a ganglion in that it neither transilluminates nor yields fluid on aspiration. It is most common on the volar and lateral aspects of the digit, and grows insidiously between tissue planes, and is yellow or brown in colour, of fibrous consistency and mostly multilobular.

522. **Giant cell tumour of the fingertip.** Gradually increasing in size.

523. **Secondary bony erosion (arrow).**

524. **Operative finding.** The tumour presented on the dorsal aspect but has grown around the lateral aspect to the pulp.

522

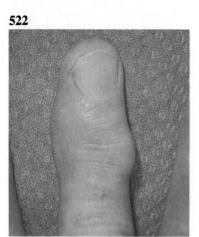

523

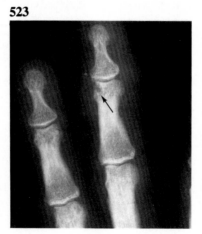

524

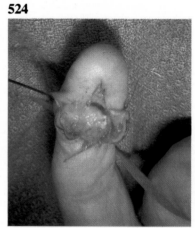

525

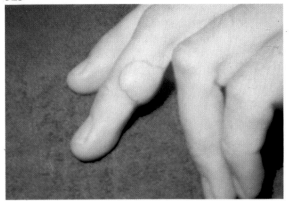

525. Giant cell tumour on the lateral aspect of a digit.

526

526. Operative finding.

527a

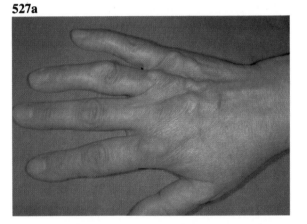

527a. Xanthomas. These are lipid deposits in the skin and tendon sheath associated with an increase in serum cholesterol. They may be familial. Multiple xanthomas are seen here on the dorsum of the hand.

527b

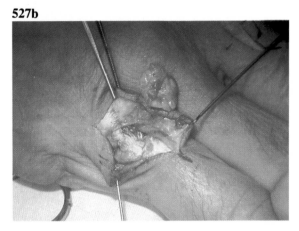

527b. Operative findings. Yellowish brown tumour of the tendon sheath resembling a giant cell tumour.

528a

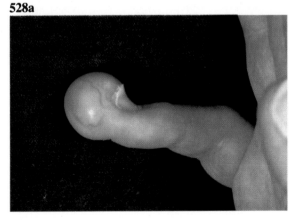

528b

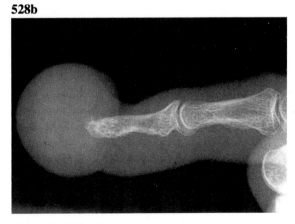

Muscle

Leiomyoma is a rare benign tumour arising from the smooth muscle, usually of blood vessels.

528a & 528b. Leiomyoma, presenting as a slow growing tumour of the fingertip. Male aged 54 years.

Blood vessels

529

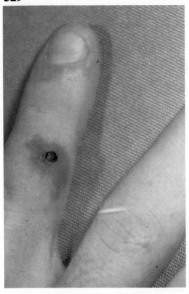

530

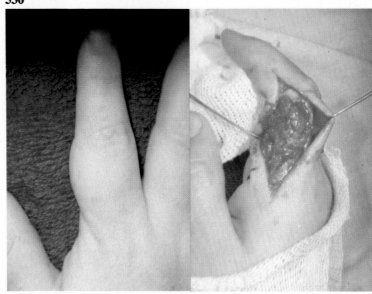

529. Capillary haemangioma, 'port wine stain'. These are flàt, pink-to-red in colour, present since birth, and do not regress with age.

530. Cavernous haemangioma, 'strawberry naevus'. A soft swelling which appears shortly after birth and grows slowly. In this case presenting as a subcutaneous tissue tumour over the proximal phalanx of the index finger of a 14-year-old school boy. On compression it empties and on release of pressure it refills. Asymptomatic, but increasing in size. The operative finding is an haemangioma infiltrating the surrounding tissue.

531 & 532. Haemangioma presenting as a small firm nodule in the palm.

531

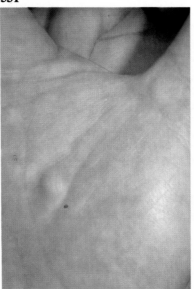

532

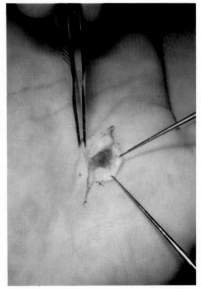

533

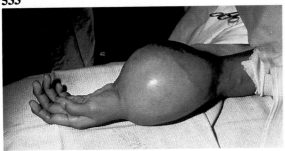

534

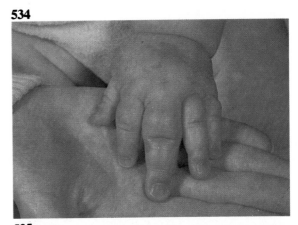

533. Haemangiolymphoma. These should always be considered in the differential diagnosis of any large or diffuse tumour.

534. Haemangiolymphoma of the middle finger, hand, and forearm.

Glomus tumour, angioneuroma, 'Popoff tumour'. This usually occurs in the region of the nail bed or the lateral pulp area. The patient presents with a severe stabbing or burning pain which can be constant or intermittent.

The pain may arise spontaneously or can be precipitated by trauma, changes in temperature or humidity. The tumour is tiny, seldom exceeding 1cm in diameter, bluish or reddish in colour and encapsulated. It can erode the distal phalanx or present as a clear punched-out defect.

Differential diagnosis. Neuroma, neurofibroma.

535a, 535b & 535c. Glomus tumour, presenting as pain at the fingertip.

535a

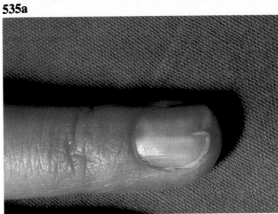

535b

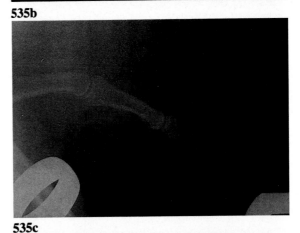

535c

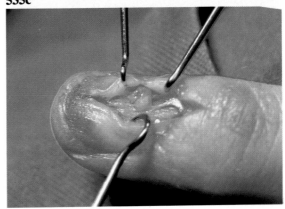

536. Glomus tumour, presenting as a painful tumour on the dorsum of the hand.

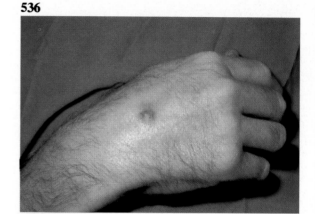

536

Nerves

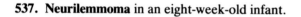

Neurofibroma. This tumour arises from the schwann cell and is of two main types – neurilemmoma (schwannoma) or neuro-fibromatosis (Von Recklinghausen's disease). A neurilemmoma is a well encapsulated fibroma of perineural tissue usually forming a single ovoid firm mass.

537. Neurilemmoma in an eight-week-old infant.

538. Neurofibroma, probably of occupational origin, on the dorsal digital nerve of a baker who persistently rubbed this part of his finger against the baking dish. A firm swelling. Operative findings, see **829**.

539. Neurofibromatosis. The enlarged fingertip is due to bone involvement.

Friedrich Von Recklinghausen (1833–1910). German pathologist. Described multiple fibromas along peripheral nerves, often associated with naevi, in 1882.

537

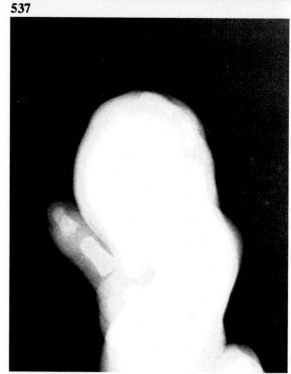

538

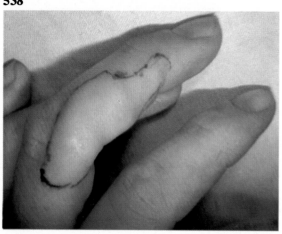

539

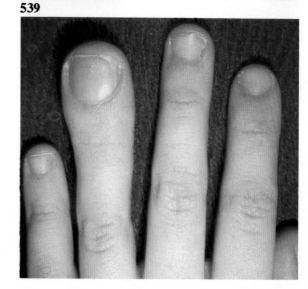

540 & 541. Neurofibrolipoma in a 22-year-old dentist. He had a firm swelling proximal and distal to the carpal tunnel in association with a carpal tunnel syndrome. The operative findings were infiltration of the median nerve with fibro-fatty tissue.

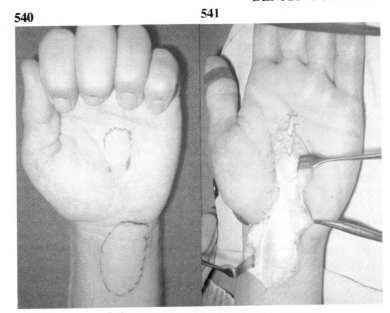

542. Haemangioma of the median nerve. Note the dumb-bell shaped swelling due to the constricting effect of the flexor retinaculum.

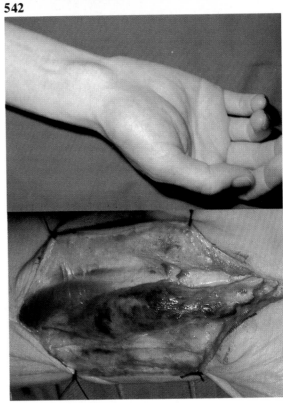

Joints

Giant cell tumours can also arise from joint synovium.

Bone

Enchondroma. 90% of bone tumours of the hand are either solitary or multiple enchondromas (Ollier's disease). About 75% present as pathological fractures, and about 20% are discovered by chance. The solitary lesions are usually found in the proximal phalanx with the adjacent middle phalanx and metacarpal the next most frequent sites. They do not occur in the carpus. Multiple enchondromas may be accompanied by multiple haemangiomas of the skin (Maffuci's syndrome).

543 & 544. Solitary enchondroma in left index finger, x-ray. Note the margin of thinned sclerotic bone over it and the tiny patches of calcified stippling within the central lytic area. Some patients present with a pathological fracture.

Differential diagnosis. Xanthoma, bone cyst.

543

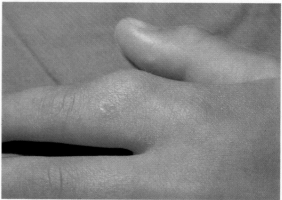

544

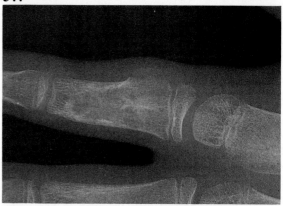

Osteochondroma (ecchondroma). This is the second most common skeletal tumour found in the hand. About 90% are discovered by the patient, who feels a hard mass, in the 2nd and 3rd decade. Their distribution is the same as an enchondroma. They can be sessile or pedunculated. When multiple they are usually associated with hereditary exostoses.

545 & 546. A subungual osteochondroma of the fingertip.

545

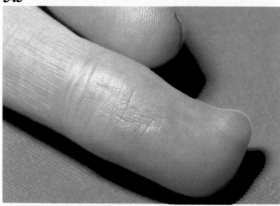

546

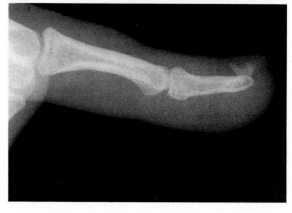

Louis Ollier (1830–1900). French orthopaedic surgeon. Described Ollier's disease, i.e. dyschondroplasia in 1899.

Angelo Maffucci (1845–1903). Italian physician. Maffucci's syndrome, chondrodysplasia (angiomatosis syndrome), is a combination of some features of Ollier's syndrome with multiple haemangiomas and phleboliths.

Aneurysmal bone cyst and giant cell tumour of bone. Both lesions have similar histories, findings and treatment. Often there is minor trauma followed by rapid enlargement of the metacarpal or proximal phalanx. X-rays show a greatly expanded thin cortex with a lacy network through the cyst.

The aneurysmal bone cyst is very vascular and very soft. The giant cell tumour, though equally vascular, is firm.

547. Giant cell tumour presenting as a pathological fracture following injury to the fingertip in a 45-year-old male.

548. Giant cell tumour, x-ray finding.

549 & 550. Giant cell tumour, six months later. The patient had refused treatment. The tumour is growing slowly.

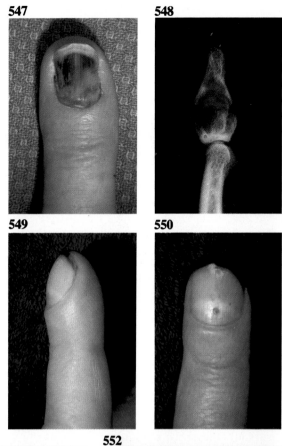

547

548

549

550

551. Giant cell tumour of the distal radius. This 21-year-old male was thought to have a sprained wrist. Persistent mild pain after a fall and slight swelling (arrow) at the wrist.

552. Giant cell tumour. X-ray.

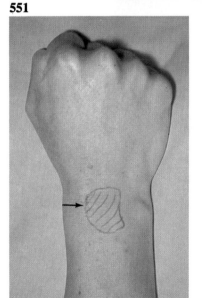

551

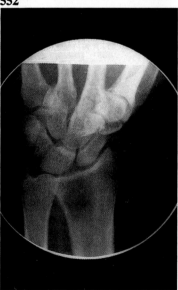

552

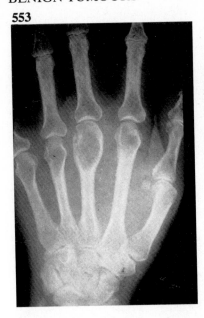

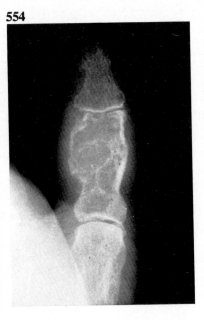

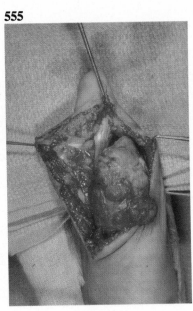

553. Aneurysmal bone cyst of the third metacarpal head.

554. Aneurysmal bone cyst of the proximal phalanx of the thumb.

555. Operative findings.

Osteoid osteoma. This rare tumour may involve any bone of the hand in the 10–38 years age group. The patient presents with a long history of aching pain which is greater at night, increased by heat and relieved by salicylates. There may be an associated tender swelling of the surrounding soft tissue.

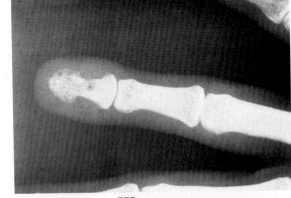

556. Osteoid osteoma in a 25-year-old waterside worker who noticed, after injury, first pain, and then a swelling of his right ring fingertip. The x-ray shows a small nidus of rarefaction. The central sclerotic focus not easily seen.

Differential diagnosis. Infection, tumour.

557. Periosteal chondroma.

558. Benign cyst in the scaphoid bone. Asymptomatic. Incidenta' finding on x-ray.

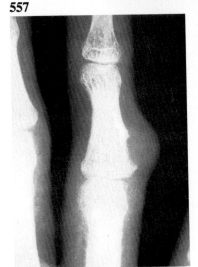

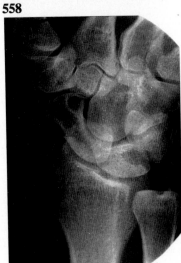

Malignant tumours

Skin

559. Bowen's disease (carcinoma-in-situ). This appears as a verrucous plaque on the hand or finger, and may proceed to squamous cell carcinoma. Patients with Bowen's disease have a predilection to skin malignancies and 25% are said to develop internal malignancies.

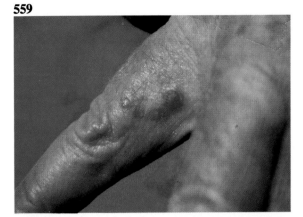

560. Basal cell carcinoma (rodent ulcer). Presenting as an unhealed sore. Mostly found on the dorsum of the hand. Begins as a pearly nodule with tiny venules coursing across its surface. Later forms an ulcer with raised, rolled, undermined and pearly edges.

Differential diagnosis. Squamous cell carcinoma.

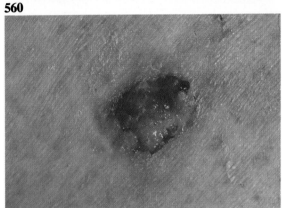

Squamous cell carcinoma. Usually begins as a slight thickening, a plaque or a small nodule. It later ulcerates. The edges of this ulcer become raised, rolled and everted. The tumour is markedly indurated and mostly found on the dorsum of the hand but it can occur in the palm.

561. Squamous cell carcinoma. Ulcer and plaque in a 76-year-old farmer.

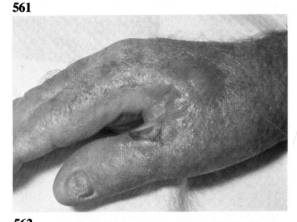

562. Squamous cell carcinoma misdiagnosed as infection and treated by ointments.

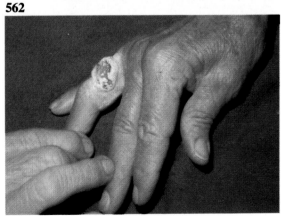

John Templeton Bowen (1857–1941). American dermatologist. Bowen's disease, a form of cutaneous epithelioma manifested as a solitary crust-covered patch over a dull, red, moist, and granular undersurface. Histologically it is a form of squamous cell epithelioma in situ.

563. Squamous cell carcinoma in a 75-year-old female.

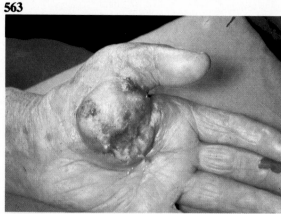

563

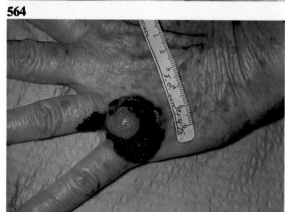

564

Malignant melanoma.

564. Malignant melanoma of the skin of the hand. It is rare but can occur on the palm or on the dorsum of the hand and presents as a growing tumour, usually flat at first and then raised. Its growth is measured in months. The lesion may be black or amelanotic. Other symptoms are bleeding and less frequently itching.

565

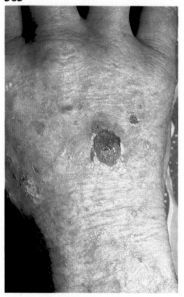

565. Amelanotic melanoma.

566

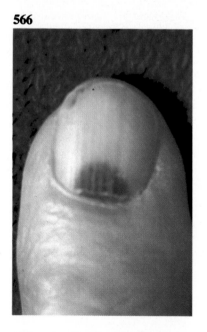

566. Subungual pigmentation can be due to melanin or haemosiderin. This was an old sub-ungual haematoma.

567–569. Subungual melanoma. This is a rare tumour and its growth may be slow, i.e. detectable over years but more commonly over months. Beware of dark lesions about the nails that do not respond to treatment. If in doubt, these lesions should be biopsied.

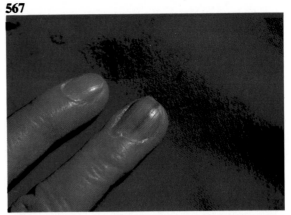

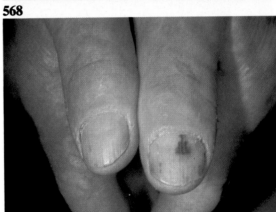

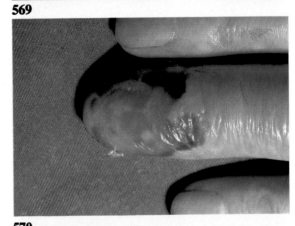

570. Amelanotic melanoma. Diagnosed by biopsy of a non-healing ulcer after crush injury to thumb.

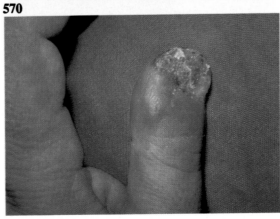

Bone

571 & 572. Osteogenic sarcoma arising from the carpus in a 76-year-old farmer.

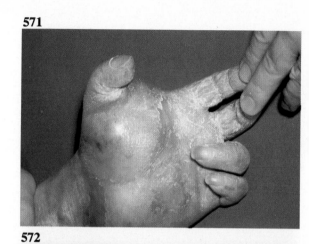

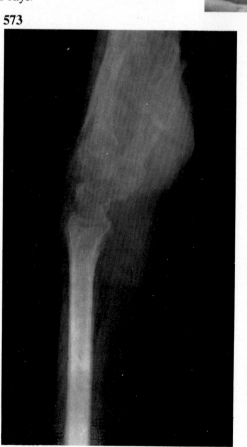

573 & 574. Osteolysis of the triquetrum and a soft tissue mass. X-rays.

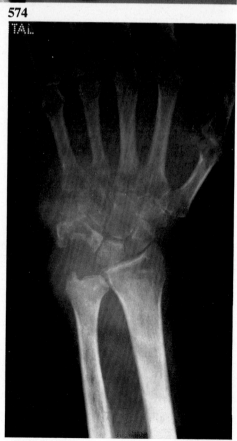

575. Osteogenic sarcoma arising from the middle phalanx of a digit. X-ray showing new bone formation and a soft tissue mass.

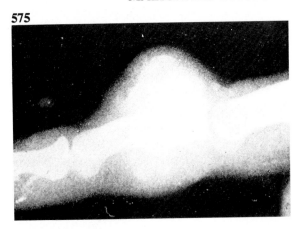

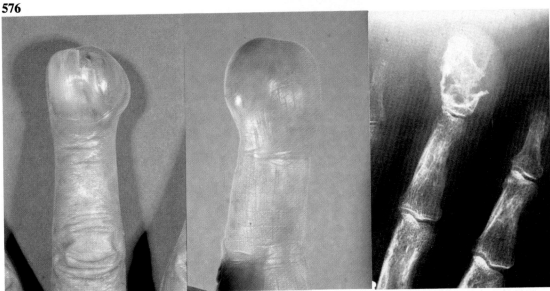

576. Chondrosarcoma. A previously stationary swelling increased in size and became painful. The x-ray shows a bone destructive tumour mass with a floculant calcification.

577. Ewing's tumour. This tumour of the medullary cavity rarely arises in the hand. It is associated with fever, local heat and tenderness and x-ray findings of bone erosion.

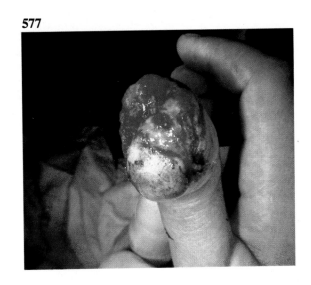

Metastatic tumours

The hand is a rare focus of metastasis from malignancies elsewhere in the body, but the primary sites are carcinoma of the lung, breast, and parotid. The terminal phalanx is most frequently involved. X-rays show an osteolytic area which does not cross the articular line.

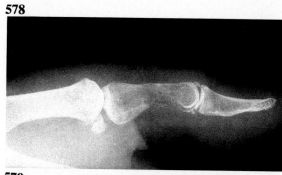

578. Metastatic tumour in the thumb of a 73-year-old male who presented with a tender red swelling of his thumb after a fall.

Rare

FAT

Liposarcomas are rare tumours. They become evident on recurrence of a previously excised 'lipoma'.

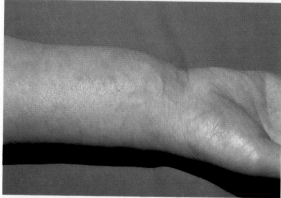

579 & 580. Liposarcoma in a 65-year-old female, presenting as a painless enlarging tumour of the distal forearm.

BLOOD VESSELS

Haemangiosarcoma is an exceedingly malignant tumour with rapid growth and early metastasis.

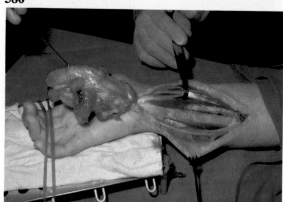

581. Haemangiosarcoma. The previous excision area was covered with a skin flap. The recurrence is growing rapidly and metastasizing.

Other rare tumours occur in:

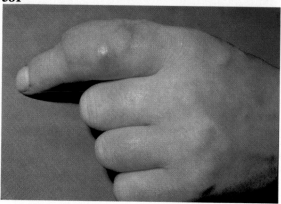

FASCIA. Fibrosarcoma is rare but manifests as a superficial, slow growing, hard, fixed mass.
TENDON AND TENDON SHEATH. Malignant synoviomas.
MUSCLE. Malignant rhabdomyosarcoma.
NERVES. A malignant neurilemmoma or malignant schwannoma arises from a pre-existing neurofibroma, with almost half of those reported occurring in patients with Von Recklinghausen's disease.
JOINTS. Synovial sarcomas have slow growth, absence of symptoms and are therefore difficult to diagnose.

6. Contractures and deformities

Diagnosing the cause of a contracture – aim

1. Which tissue is involved?	Skin (scar, burn, scleroderma) Fascia (Dupuytren's contracture) Muscle (ischaemic fibrosis, nerve palsy or spasticity) Tendon (adhesions, division) Nerve (palsy or spasticity) Joint (congenital contracture, traumatic or infective fibrosis, arthritis) Bone (malunited fracture)
2. What is the pathological process?	Congenital (camptodactyly) Injury (scar contracture, nerve or tendon division, adhesions) Infection (scar contracture) Ischaemia (Volkmann's contracture) Degenerative (arthritis, Dupuytren's contracture)
3. Is it functional?	Hysterical contracture

Diagnosing the cause of a contracture – method

1. History	Injury, burn, infection, ischaemia Prolonged immobilisation Family history, e.g. Dupuytren's contracture
2. Examination	COMPARE BOTH HANDS *Look for* – scars (tendon injury, nerve injury) – fascial bands, nodules (Dupuytren's contracture) – skeletal deformity (fracture, dislocation) *Test the contracted joint for* – active movements (absent active movements of a passively mobile joint means the neuromuscular-ligamentous mechanism is at fault) – passive movements (if absent, the joint is contracted) *Alter the position of the more proximal joint and repeat the test* (see the Appendix, page 341)
3. Investigation	To show skeletal lesions, x-ray the relevant bones and joints in true antero-posterior, lateral and oblique positions

MUSCLE-TENDON FORCES

PRIME ACTION

radius and ulna
metacarpal
proximal phalanx
carpus

wrist

flexion, extension and stabilisation

wrist extensors
wrist flexors
middle phalanx
distal phalanx

finger

proximal interphalangeal joint flexion

distal interphalangeal joint flexion

flexor digitorum superficialis

flexor digitorum profundus

extensor digitorum

metacarpo-phalangeal joint flexion

metacarpo-phalangeal joint extension

interphalangeal joint extension

intrinsics
(interossei and lumbricals)

unopposed action of extensor digitorum → hyperextension of metacarpo-phalangeal joints

unopposed action of flexors (flexor digitorum profundus and superficialis) → flexion of interphalangeal joints

extensor digitorum

flexor digitorum profundus and superficialis

paralysed intrinsics

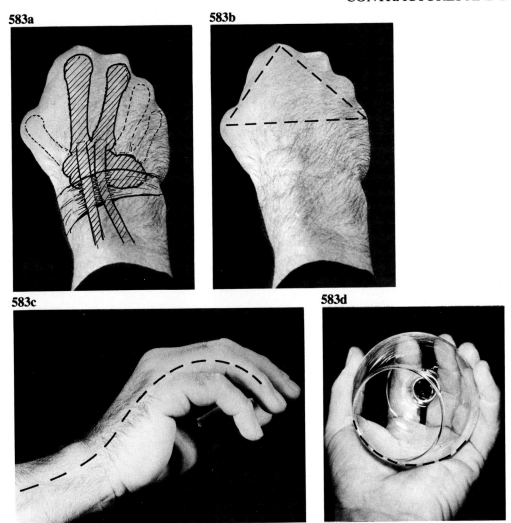

583a

583b

583c

583d

582. The balance of muscle tendon forces acting on the hand.

Hand function is dependent on a finely balanced and integrated system of synergistic and antagonistic muscle-tendon forces acting on a long cantilevered skeletal framework.

Twenty-eight *extrinsic* muscles arise in the forearm. They flex, extend, and stabilize the wrist, extend the metacarpo-phalangeal joints and flex the interphalangeal joints. They provide the hand with power, but their function is dependent on the balancing action of the intrinsic muscles.

Twenty *intrinsic* muscles arise from the hand itself. They position the thumb in a plane at right angles to the hand. They flex, abduct, and adduct the metacarpo-phalangeal joints and extend the interphalangeal joints. They provide dexterity and fine movements.

583a, b, c, d. Fixed unit. Backbone of the hand. Arches. The five rays of the skeletal frame do not lie in a flat plane. The metacarpal heads present a 'triangle' (**b**) with the 1st and 5th forming the base and the index representing the apex.

The hand comprises a central fixed unit and two mobile lateral units (**a**). The 2nd and 3rd metacarpals provide the backbone or fixed unit (shaded). This is wedged into the fixed distal carpal arch and is motivated by the powerful radial wrist extensors and flexor. (E.C.R.L. & B, F.C.R.)

The longitudinal arch (**c**) is maintained by the balanced action of the intrinsics and extrinsics, see **582**.

The transverse arch (**d**) is maintained by the thenar and hypo-thenar intrinsics, see **18**.

HAND POSTURES AND POSITIONS

The posture of each joint in the hand is dependent on the balance of those muscle forces acting on it. This balance is controlled by the posture of the joint proximal to it (page 184). The wrist is the key joint of the hand, controlling as it does the metacarpo-phalangeal and interphalangeal joints beyond it.

584. The position of function. When the hand is relaxed and the forearm is supported in supination, the weight of the hand causes the wrist to fall into dorsi-flexion. The fingers lie in increased flexion at each joint, from index to little finger. The thumb lies abducted and slightly flexed, with its pulp opposed to the index pulp.

585. The position of injury or rest. When the forearm is pronated, the wrist falls into flexion. The fingers and thumb now become extended.

Definitions

Posture. Place, position or situation. The situation or disposition of several parts of the body with respect to each other or for a particular purpose.

Position. Attitude, condition, or manner in which anything is placed – site, place, station. The spot where a thing is placed.

586. Diagnosing the cause of a joint contracture by altering the posture of the more proximal joint.

See Appendix, page 341.

584

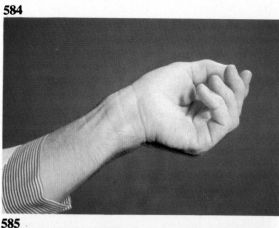

585

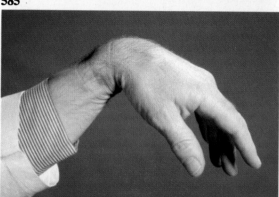

CLASSIFICATION OF CONTRACTURES

Contractures of the hand may be classified as flexor, extensor or web (adduction contractures). They may be reversible or irreversible, i.e. fixed.

Although the disorder may begin in one tissue such as the skin, secondary changes can occur in adjacent tissues such as the fascia, tendon or joint.

When hand joints are injured they assume a position in which their ligaments are most relaxed. The ligaments subsequently fibrose and a joint contracture develops.

Characteristically the metacarpo-phalangeal joints stiffen in extension and the interphalangeal joints stiffen in flexion (see page 341).

587 & 588. Claw hand deformity, following crush injury. Caused by the relaxation and subsequent contraction of the collateral ligaments of the metacarpo-phalangeal joint, and the palmar plate of the interphalangeal joint.

587

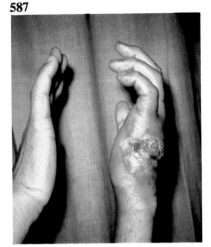

588

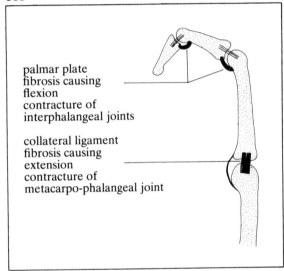

palmar plate fibrosis causing flexion contracture of interphalangeal joints

collateral ligament fibrosis causing extension contracture of metacarpo-phalangeal joint

Skin

Cutaneous scars crossing the flexion creases at right angles tend to contract. This occurs most frequently in the palm and web of the young, and in individuals predisposed to excess collagen deposition.

589. Flexion contracture of a child's right little finger from a skin laceration crossing the flexion creases of both interphalangeal joints.

589

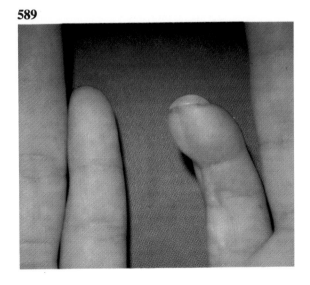

590. Adduction contracture of the thumb. Thumb abduction and extension is prevented by the scar which crosses the thenar crease.

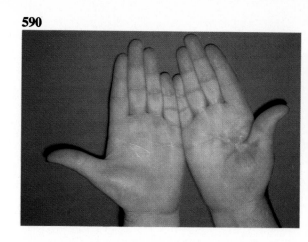

590

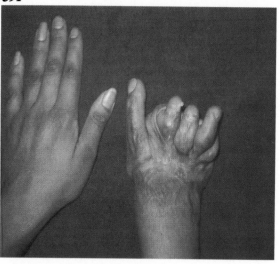

591

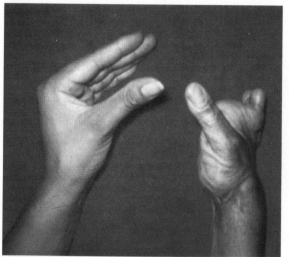

592

591–593. Extension contracture in the right hand of a 25-year-old Botswana female from dorsal burns incurred in childhood. The metacarpo-phalangeal joints are fixed in hyperextension and the interphalangeal joints in flexion. There was secondary fibrosis of the extensor tendons. Fibrosis of all tissues in the first web has caused adduction contracture of the thumb. The severity of the skeletal deformity is seen in the x-ray.

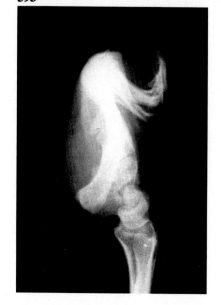

593

594. Flexion contracture. This 30-year-old labourer had the palmar surface of his right ring and little fingers burnt as a child. Without treatment the scars contracted, causing this irreversible flexion contracture.

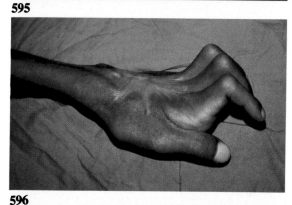

595. Extension contracture. This Indian had extensive anthrax infection on the dorsum of his hand. Healing was complicated by fibrosis of the extensor tendons and metacarpo-phalangeal joints.

Fixed hyperextension of the metacarpo-phalangeal joints inhibit the extensor action of the intrinsics on the interphalangeal joints. Unopposed, the flexor profundus and superficialis cause a flexion contracture of the interphalangeal joints.

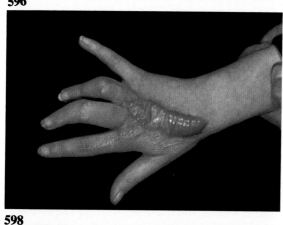

596. Keloid, causing extension contracture of the middle and ring fingers.

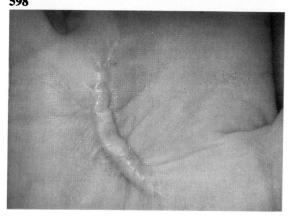

597 & 598. Keloid from a burn, causing adduction contracture of the thumb. A keloid is a raised, red, irritable proliferation of fibroblasts in the wound. It is more than scar hypertrophy; it tends to recur after excision.

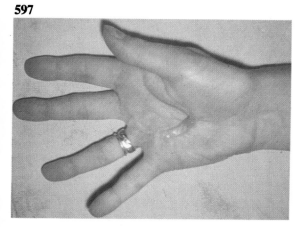

599. Scleroderma. Tightening of the skin and sub-cutaneous tissues of the fingers may result in a stiff claw deformity. Subcutaneous calcification, osteoporosis, and destruction of the distal phalanges may accompany the connective tissue sclerosis.

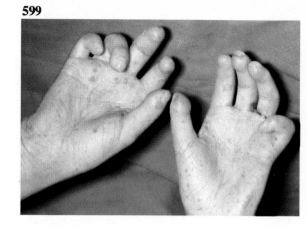

599

Fascia

Dupuytren's contracture. Contracture of the palmar fascial bands produces flexion contracture of the metacarpo-phalangeal and proximal inter-phalangeal joints, the flexor tendon apparatus and the skin itself.

When injury or operation is associated with severe pain some patients hold their hand and fingers in a flexed position. The palmar fascia may fibrose and contract and produce a secondary flexion contracture of the metacarpo-phalangeal joints, as in Dupuytren's contracture.

600. Dupuytren's contracture – a later stage. Flexion contracture of the proximal inter-phalangeal joint of the right middle finger and of all three joints of the left little finger.

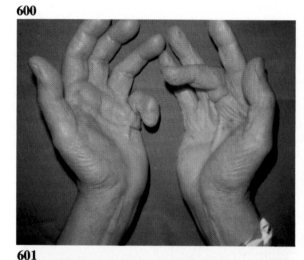

600

601

601. Post-traumatic palmar fasciitis, causing flexion contracture of all fingers. This condition followed three months after a small laceration in the palm. Because of pain, this patient kept her fingers cupped.

Baron Guillaume Dupuytren (1777–1835). French surgeon. Most renowned surgeon in Europe in his time. Reported palmar fascial contracture in 1833, although this was first described by Astley Paston Cooper in 1822.

Tendon

Adhesions

For a muscle to affect joint action, its tendon must glide either through a synovial tube and pulley system or through loose paratenon tissue.

After injury, operation or infection, adhesions may form and restrict this tendon glide.

602. Extension contracture of the little finger from scar (arrow), involving the skin and extensor tendons of the little finger. This 16-year-old boy had poor grip from stiffness of the metacarpophalangeal joint of his little finger.

603. Flexor contracture of the right middle finger from adhesions developing after flexor tendon repair. The finger will not passively extend.

604. Operative findings. Adhesions binding the flexor tendon to its fibrous tunnel.

605. Absent dorsal skin creases from failure of joint movement.

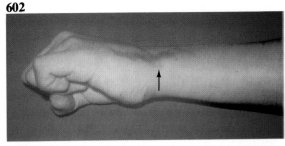

602

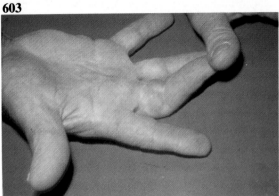

603

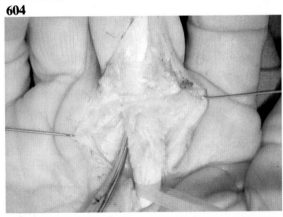

604

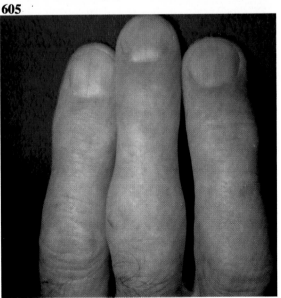

605

Ruptures

606. Types of finger deformity from muscle tendon imbalance.

606

TYPE	MECHANISM			
The normal balance of muscle tendon forces	*joint*	*flexors*	*extensors*	
	MCPJ	intrinsics	EDC	
	PIPJ	FDS, FDP	intrinsics (MET)	
	DIPJ	FDP	intrinsics (TET)	

Ulnar claw hand	Paralysis of intrinsic flexion of MCPJ and extension of IPJ →
	EDC hyperextends MCPJ
	FDP and FDS flex IPJ

Swan neck (recurvatum deformity)	Hyperextension of PIPJ from either:
	1. weak flexion – division of FDS, or
	2. strong extension (a) tight intrinsics from rheumatoid arthritis or ischaemic fibrosis
	(b) division of TET and retraction of LET

Boutonnière (button hole deformity)	Division of MET → flexion PIPJ from unopposed action of FDS.
	The LETs prolapse to the palmar side and further flex PIPJ and hyperextend DIPJ.

Mallet finger	Division of TET → flexion DIPJ from unopposed action of FDP.
	(Retraction of LET → hyperextension PIPJ)

607

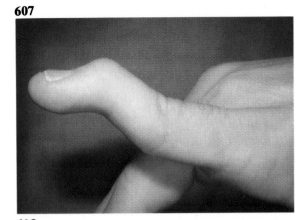

607. Mallet finger. Flexion contracture of the distal interphalangeal joint from closed rupture of the terminal extensor tendon. The lateral extensor tendons (lateral bands) displace dorsally, causing secondary hyperextension of the proximal interphalangeal joint.

608

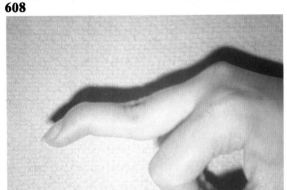

608. Rheumatoid intrinsic muscle fibrosis can cause a similar contracture. Fibrosis of the intrinsic extensor mechanism over the proximal interphalangeal joint causes a secondary extensor contracture of that joint, and a flexion contracture of the distal interphalangeal joint from the unopposed action of flexor digitorum profundus.

609

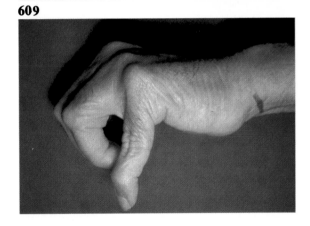

609. Swan neck deformity from division of flexor digitorum superficialis. Absence of the flexor digitorum superficialis stabilising force on the proximal interphalangeal joint alters the intrinsic extrinsic balance. The unopposed lateral bands now hyperextend the proximal interphalangeal joint and flexor digitorum profundus flexes the distal interphalangeal joint.

610. Boutonnière deformity from division of the medial extensor tendon ('central slip'). Irreversible joint contracture resulted from a tiny cut over the proximal interphalangeal joint one year before.

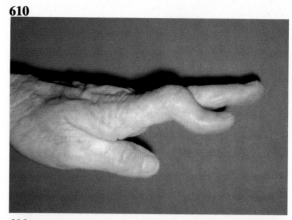

611. Bony ankylosis.

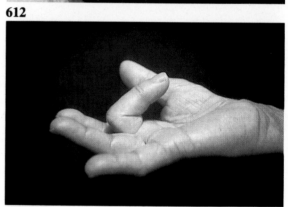

612. Trigger finger. Entrapment of the flexor tendon within the pulley at the base of the fingers causing a flexion contracture.

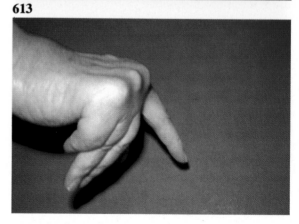

613. Subluxation of extensor tendons with flexion contracture of the metacarpo-phalangeal joints. Rupture of the extensor hood allows the extensor tendons to slip from the metacarpal head to the intermetacarpal groove, where they flex rather than extend the metacarpo-phalangeal joint.

Muscles

Ischaemia

Volkmann's contracture, i.e. ischaemic contracture of the flexor muscle mass can follow brachial artery injury, burns, crush oedema, and congestion from a tight plaster.

The five 'P's' of impending Volkmann's ischaemic contracture are:

pain
pallor
paralysis
paraesthesia
pulselessness

Richard Von Volkmann (1830–1889). One of the best known German surgeons of the mid-19th century described ischaemic contracture in 1881.

614 & 615. Volkmann's contracture. Three months after a deep forearm burn. Fibrosis and shortening of the flexor muscle tendon units produced a flexion contracture of the fingers. This is aggravated by extension of, and partially relieved by flexion of, the wrist.

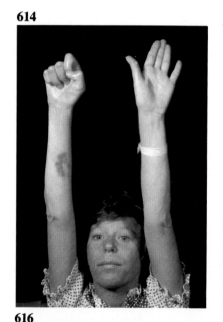

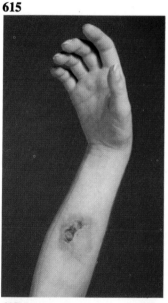

616. Operative findings. Ischaemic fibrosis and necrosis of the flexor muscle.

617. Supracondylar fracture of the humerus, a dangerous fracture, which may lead to brachial artery damage, acute ischaemia, and Volkmann's contracture.

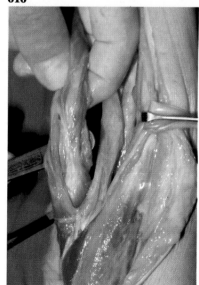

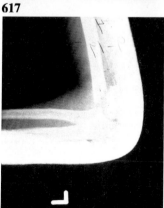

618

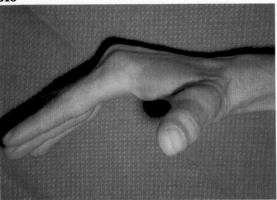

618. Intrinsic muscle contracture developing three months after crush injury. In this, the 'intrinsic plus' position, there is flexion of the metacarpo-phalangeal joints and extension of the inter-phalangeal joints from ischaemic fibrosis of the intrinsic muscles. This patient cannot flex the inter-phalangeal joints.

619

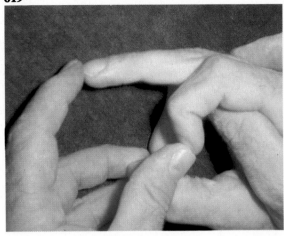

619. Testing for intrinsic tightness. Any tightness of the intrinsic muscle tendon unit will be accentuated by extension of the metacarpo-phalangeal joint. In this patient who had intrinsic contracture of his middle finger, extension of the metacarpo-phalangeal joint prevented passive flexion of the interphalangeal joints. The index and little fingers showed normal passive flexion.

Flexion of the metacarpo-phalangeal joint of the middle finger relaxed the intrinsics sufficient to allow some passive flexion of the interphalangeal joints.

620

620. Intrinsic contracture of the left ring finger from post-operative palmar haematoma, with subsequent ischaemia and fibrosis of the intrinsic muscles of that finger. Note the flexion contracture of the metacarpo-phalangeal joint, hyperextension of the proximal interphalangeal joint, and flexion deformity of the distal interphalangeal joint.

621

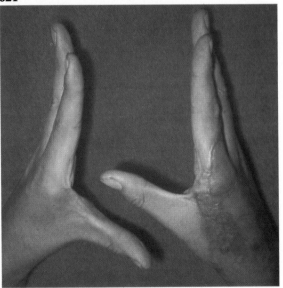

621. Thumb web contracture from traumatic ischaemic fibrosis. There is a skin graft on the dorsum of the hand.

Nerve palsy and spasticity

The normal position of the hand depends on normal muscle tone, and this of course depends in turn on a normal nerve supply to that muscle. Nerve palsy and spasticity can cause muscle imbalance and joint contractures can subsequently occur.

622. Ulnar claw deformity of the left hand from intrinsic muscle palsy. This woman had ulnar nerve injury at the wrist. Paralysis of the ulnar innervated intrinsic muscle leads to hyperextension of the metacarpo-phalangeal joint from the unopposed action of extensor digitorum, and flexion of the interphalangeal joints from the unopposed action of the extrinsic flexors.

623. Flexion contracture of the metacarpo-phalangeal joints of the ring and little fingers. Partial injury to the radial nerve in the upper forearm paralysed the muscle tendon units to these fingers.

624. Spastic right hand after cervical cord injury. This patient could not actively or passively extend his fingers or thumb.

622

623

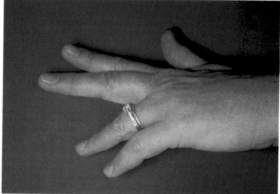

624

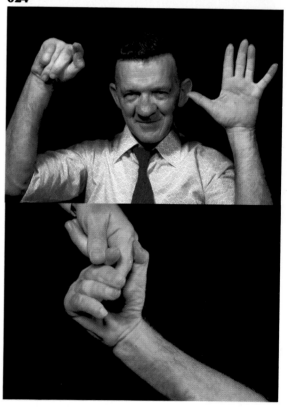

Joints

Congenital contractures

625

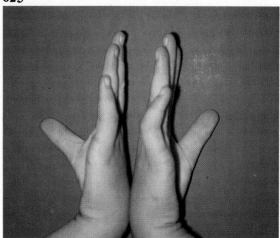

625. Camptodactyly. Congenital flexion contracture of the proximal interphalangeal joints of each little finger. See also **126**.

626. Clinodactyly. Lateral and flexion deviation of the proximal interphalangeal joint. See also **132– 134**.

627. Congenital flexion contracture of both interphalangeal joints of the right little finger and of the proximal interphalangeal joints of the right ring and middle fingers in a 31-year-old fisherman.

626

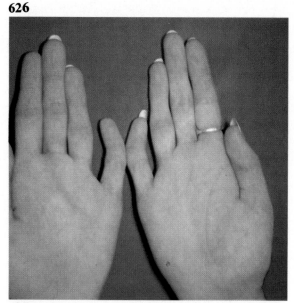

627

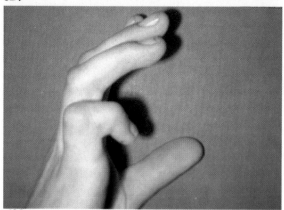

628

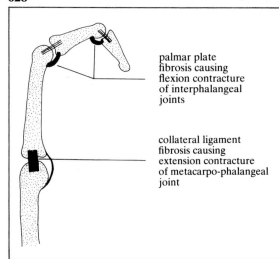

Acquired contractures

When hand joints are injured or immobilised for prolonged periods, they assume a position in which their ligaments are most relaxed. The ligaments subsequently contract, causing the joints to become rigid in their relaxed position. The characteristic relaxed position for various joints of the hand is as follows:

wrist – *flexion*
thumb – *adduction*
metacarpo-phalangeal joints – *extension*
interphalangeal joints – *flexion*

628. Position of relaxation and subsequent contraction of the ligaments of the metacarpo-phalangeal and interphalangeal joints. See **588**.

palmar plate fibrosis causing flexion contracture of interphalangeal joints

collateral ligament fibrosis causing extension contracture of metacarpo-phalangeal joint

629. **Contracture of the left thumb** from long-standing osteo-arthritis and subluxation of the carpometacarpal joint. Secondary muscle imbalance causes hyperextension of the metacarpophalangeal joint and flexion of the interphalangeal joint.

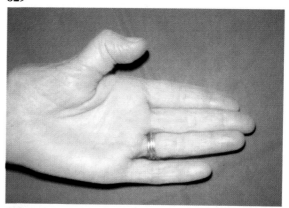

630. **Osteo-arthritis** and subluxation of the carpometacarpal joint and deformity of the metacarpophalangeal and interphalangeal joint.

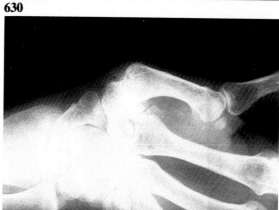

631. **Flexion contracture** of the proximal interphalangeal joint from injury. When injured the proximal interphalangeal joint assumes the flexed position, and the metacarpo-phalangeal joint assumes the extended position.

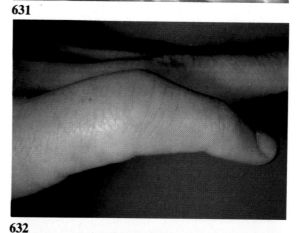

632. **Irreversible contracture** of the proximal interphalangeal joint of the right little finger from scarring, which first affected the skin, but later affected all articular and peri-articular tissues.

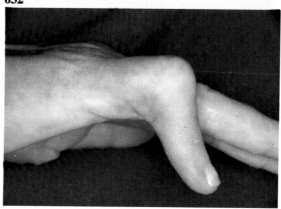

633. Flexion adduction contracture of the thumb from septic arthritis of the metacarpo-phalangeal joint.

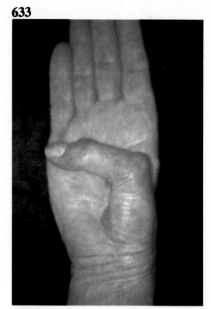

634. Flexion contracture from rheumatoid arthritis. Dislocation of the metacarpo-phalangeal joint of the left index finger.

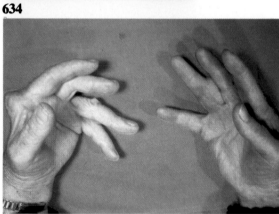

635. Flexion contracture of the proximal inter-phalangeal joints. Iatrogenic. Immobilisation in the wrong position for too long. The patient, a 52-year-old typist, was treated in a plaster cast for her sprained wrist. She had not injured her fingers. However, the interphalangeal joints were held in flexion in the plaster for six weeks. She won a law suit against her doctor.

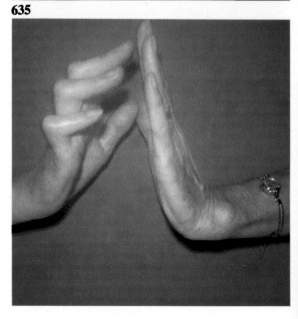

Bones

636. Muscle tendon forces causing angulation of a fractured proximal phalanx. The extensor digitorum and the lumbrical pull the head of the proximal phalanx dorsally. The interosseous pulls the base of the proximal phalanx volarly. Once started, the angulation is self-perpetuating.

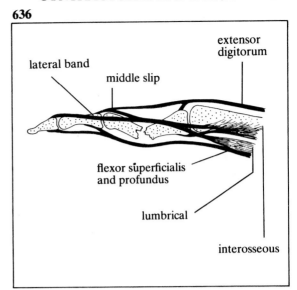

637. Flexion contracture of the proximal interphalangeal joint of the right index finger. Malunion of the fractured proximal phalanx has caused adhesions to the intrinsic mechanism. The line of action of the intrinsics has also been altered, preventing their full extension action on the proximal interphalangeal joint.

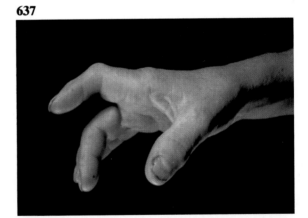

638. Malunited fracture of the proximal phalanx.

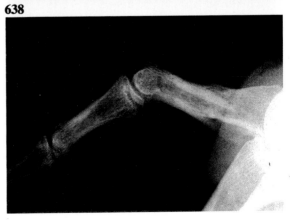

Functional contractures, hysteria

Individuals can voluntarily or involuntarily hold their hands in various bizarre positions. The contractures are caused either by intrinsic or extrinsic muscle spasm.

The diagnosis is usually obvious. Mostly there is a sudden onset of the contracture after injury or operation in a person prone to a compensation neurosis or hysteria. There may be associated hysterical anaesthesia without an anatomical basis and without trophic signs. This is called 'glove and stocking anaesthesia'. The skin texture is usually normal.

The patient can often be tricked into relaxing the spasm.

639. Functional contracture of the left dominant hand in a labourer who had a *compensation neurosis*. There is spasm of the intrinsic muscles of the thumb and fingers.

640. Functional contracture of the left middle and ring fingers. This patient had voluntary spasm of the intrinsic muscles, causing flexion of the metacarpo-phalangeal joints and extension of the interphalangeal joints.

641 & 642. Functional adduction contracture of the thumb. This man, a butcher, had many years ago amputated his left index finger. He is now seeking compensation. He holds his thumb firmly against the index finger, but when his attention is diverted his hand assumes a more normal position. There are no objective abnormal motor or sensory signs in the hand.

639

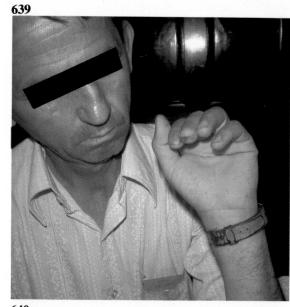

640

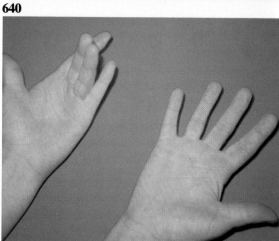

641

642

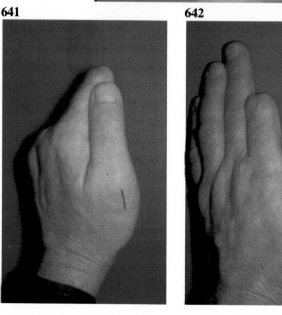

643. Irreversible flexion contracture of the little finger with secondary palmar fascial contraction. This was the end result of an *hysterical spasm*. The patient, a model, had been assaulted six months before; the only wound was a tiny cut on the pulp. She subsequently committed suicide.

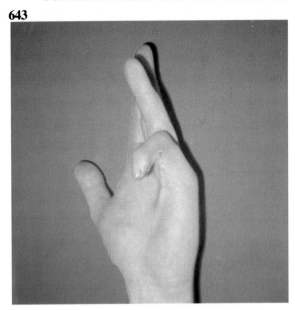

643

7. Vascular disorders

For vascular injuries see chapter 3.

Clinical features of acute arterial limb ischaemia – the five 'P's

CAUSE	FACTORS	PATHOGENESIS	CLINICAL FEATURES, 'P'

1. Arterial injury
  – open
  – closed
		Partially ischaemic muscle (totally ischaemic muscle is pain free)	PAIN

	Blood vessel
	  – anatomical site
	  – pre-existing disease
	  – collateral circulation
		Ischaemic nerve (sensory)	PARAESTHESIA

2. Arterial occlusion
  – thrombus
  – embolus
		Ischaemic muscle and nerve (motor)	PARALYSIS

	Reduction in blood flow – sudden or slow		
3. Arterial spasm		Loss of arterial flow	PALLOR & COLDNESS or mottled blue, white colour
			PULSELESSNESS

Arteries

Occlusive disorders (atherosclerosis, thrombosis, embolism)

Occlusive vascular diseases affect the vessels of the upper limb and hand as well as the lower limb and foot. The clinical picture depends on the vessels that are occluded, the speed and extent of the occlusion, and the state of the proximal and collateral circulation.

644. Arteries supplying the hand.

(1) radial artery
(2) superficial palmar branch
(3) princeps pollicis artery
(4) radialis indicis
(5) proper palmar digital artery
(6) common palmar digital artery
(7) superficial palmar arch
(8) deep palmar arch
(9) ulnar artery

644

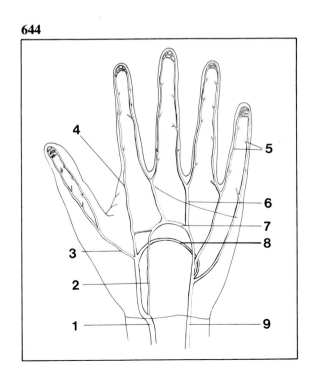

645

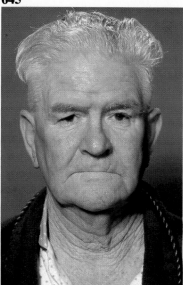

646

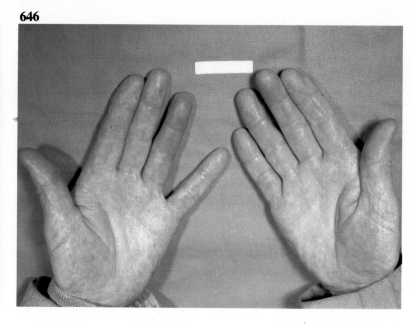

645–650. Atherosclerosis in a 75-year-old man affecting his coronary, upper and lower limb vessels.

646. Slowly progressive ischaemia of his fingers manifesting as pain, paraesthesia, and coldness, especially in his left hand. In this photograph his left ring fingertip is blue.

647

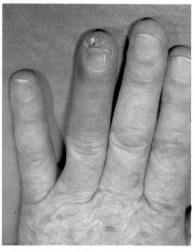

648

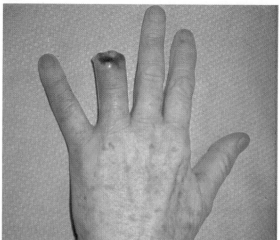

647. Recurrent infection and disordered nail growth in the left ring fingertip – from ischaemia. This finger was subsequently amputated.

648. Necrosis in the amputation stump of the left ring finger. The circulation was not adequate to heal the skin flaps, so the entire finger was later amputated.

649 & 650. Raynaud's phenomenon. Periodic episodes of painful vasospasm. There is blueness of the right middle finger pulp and pallor of the left index finger.

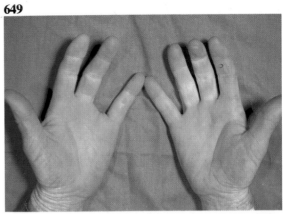

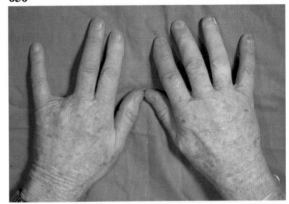

651. Thrombosis of the brachial artery and vein causing acute ischaemia of the hand. Four of the five classic signs were present – pain, paraesthesia, paresis, pulselessness. There was congestion rather than pallor because of the associated venous obstruction.

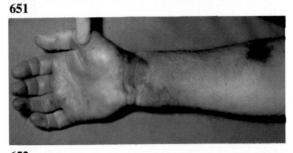

652. Axillary artery embolus with acute ischaemia of the fingers and gangrene of the right index and ring fingertips.

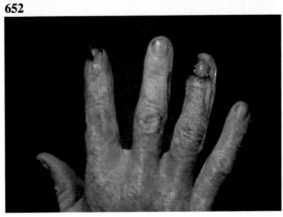

653

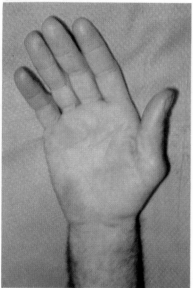

653. Ulnar artery embolus in a 45-year-old male with mitral stenosis. Acute pain and ischaemia of the right little finger. There is congestion of the finger and bruising over the hypothenar eminence.

654

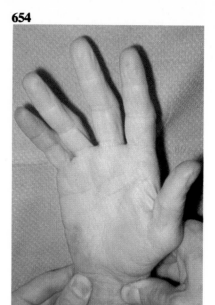

654. Feeling for the radial and ulnar pulse. The ulnar pulse could not be felt.

655

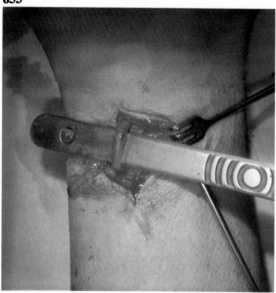

655. Operative finding. Embolus in the ulnar artery.

656

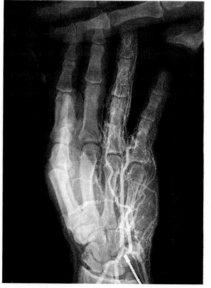

656. Ulnar arteriogram. Poor filling of the vessels supplying the little finger.

657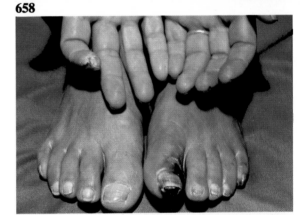

657. Digital artery embolus of the thumb with acute ischaemia.

Buerger's disease (thrombo-angiitis obliterans) is an episodic, segmental, inflammatory, and thrombotic disease affecting the arteries and superficial veins of both upper and lower limbs, usually in men between 25 and 40 years of age. It is rare in non-smokers.

The distinguishing features from atherosclerosis include: a more dramatic clinical picture, persistent rest pain, intermittent course, and a greater incidence of gangrene.

658. Buerger's disease in a 30-year-old man with gangrene of his left big toe and acute ischaemia of his right index finger.

659. Buerger's disease with gangrene of the thumb.

658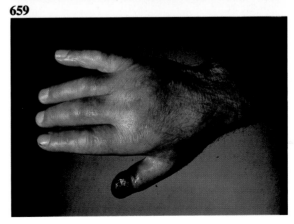

659

Leo Buerger (1879–1943). New York pathologist. Described thrombo-angiitis obliterans (inflammation of a vessel with thrombus formation) in 1908.

660. Scleroderma in a 22-year-old female. Acute on chronic fingertip ischaemia. See also **865–868**.

In scleroderma there are 'characteristic skin changes and poor circulation in the fingers. The digital ischaemia is caused by perivascular and intimal fibrosis. Clinically there is pain, numbness, coldness, and later fingertip atrophy.

Vasospastic disorders

Raynaud's disease and phenomenon. The primary Raynaud's disease or secondary Raynaud's phenomenon is characterised by intermittent attacks of pallor, followed by cyanosis in the fingers and occasionally in the toes. It is precipitated by cold or emotion. It usually begins in one or two fingertips, but later involves all the fingers, though rarely the thumb.

661. Raynaud's disease in a 27-year-old Greek woman presenting with episodes of vasospasm in the fingers of both hands. The dressing on her neck covers the wound of a right cervical sympathectomy operation.

662. An episode of vasospasm.

663. Cyanosis of all the fingertip pulps.

Maurice Raynaud (1834–1881). French physician. Described local asphyxia and symmetrical gangrene, since known as Raynaud's disease, in 1862.

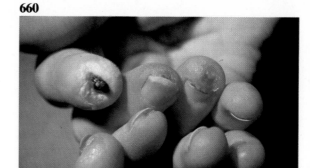

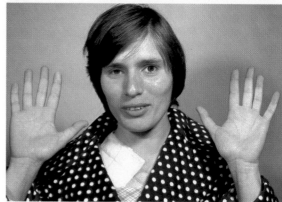

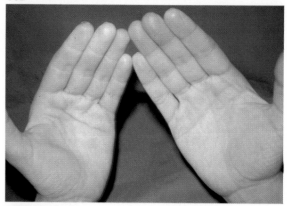

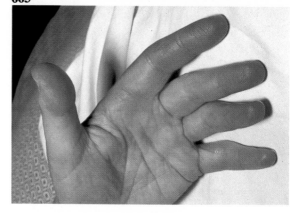

664. The recovery phase. The bright red return of colour beginning at the base of the fingers.

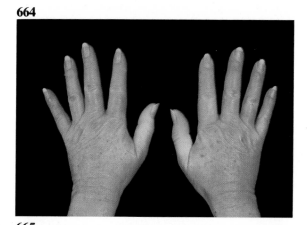

665. Raynaud's disease in a 36-year-old woman. There are small painful recurrent necrotic ulcers of the fingertips, wasting of pulps, and irregular nail growth.

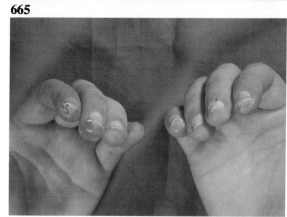

666. Raynaud's disease with vascular changes in all four limbs.

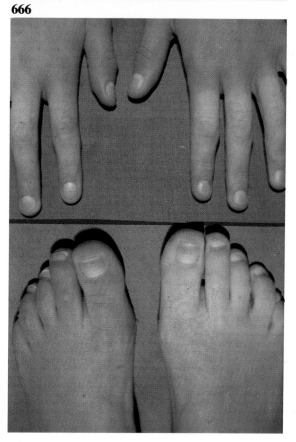

Raynaud's phenomenon may be primary, or secondary to such conditions as vibration injuries, atherosclerosis, collagen disease, malignancy, blood disorders (e.g. polycythaemia), and cervical rib syndrome.

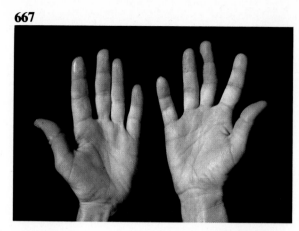

667. Raynaud's phenomenon in a patient with scleroderma. The bases of the fingers are pale and the tips congested.

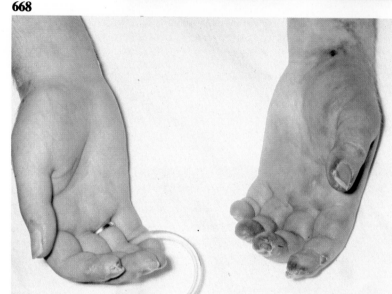

668. Raynaud's disease, if severe enough, goes to necrosis of the fingertips and even more extensive gangrene when occlusion of the small vessels supervenes on vasoconstriction. This patient had lupus erythematosis.

Vasospastic disorders can be associated with Sudeck's atrophy, causalgia, and trauma.

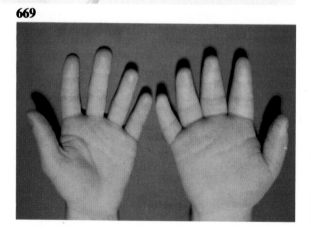

669–671. Sudeck's atrophy of the right hand developing six weeks after a Colles' fracture. Vasomotor hyperactivity can also occur after minor injury (see page 237). The patient complains of burning pain and swelling. The hand is usually warm at first, then becomes cold and cyanotic and later becomes trophic, with joint stiffness.

Paul Hermann Martin Sudeck (1866–1938). German surgeon. Sudeck's atrophy, Kienboeck's atrophy, reflex sympathetic dystrophy, reflex bone atrophy. Osteoporosis, usually of the wrists, hands, and feet in association with swelling and tenderness of underlying soft tissues, which follows fractures or minor injuries. Autonomic vasomotor disorders are the suspected cause.

Abraham Colles (1773–1843). Irish surgeon. Professor of anatomy and surgery. Described the common type of fracture of the lower end of the radius in 1814.

670

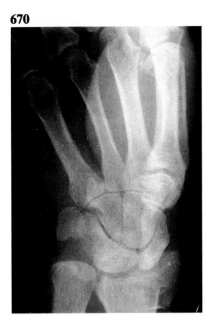

671

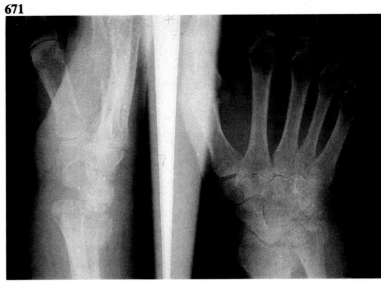

670. Colles' fracture. X-ray on the day of injury. Normal bone density.

671. X-ray at six weeks. Osteoporosis of the carpal bones.

672. Causalgia. A 35-year-old woman, whose median nerve was penetrated by a splinter, developed intense burning pain, particularly in the index finger. Amputation of this finger did not relieve the pain, which was aggravated by touch and changes of temperature. She became severely neurotic, and developed painful flexion contractures of the fingers and a trophic right thumb (red and smooth). Satisfactory function remained only in her little finger.

672

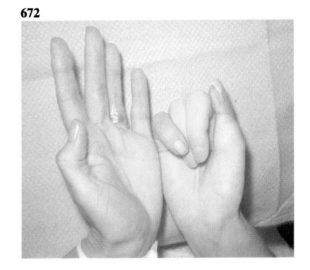

Causalgia. The Greek derivation means a burning heat. The term was first described by Silas Weir Mitchell (American neurologist, 1829–1914) in 1864. He used the term to mean burning pain associated with glossy skin.

673

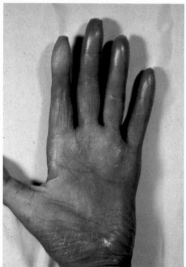

674

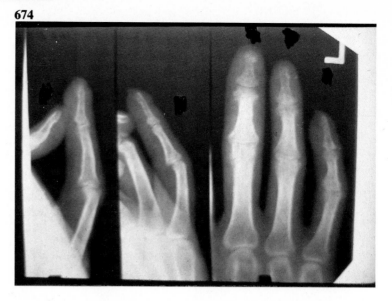

673. Post-traumatic vasomotor dystrophy occurring in the left middle, ring, and little fingers of a 40-year-old Negro. These fingers are painful, congested, and shiny.

674. Osteoporosis of the distal phalanx in these three digits.

Cervical rib. Thoracic outlet compression syndrome. Costoclavicular syndrome. The subclavian artery and lower portion of the brachial plexus can be compressed as they arch across a cervical rib or a fibrous band, extending from it to the first rib. The patient may present with an ischaemic hand, or more likely with a vasospastic condition similar to Raynaud's disease. Post-stenotic aneurysmal dilatation, embolisation and thrombosis can occur.

676

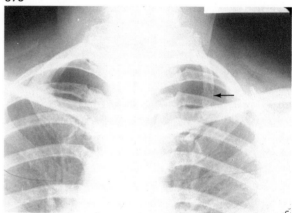

675

675. Cervical rib. Most patients with cervical ribs are asymptomatic, but this Italian housewife had Raynaud's phenomenon – intermittent attacks of pain, pallor, and cyanosis after prolonged housework.

676. X-ray of cervical rib (arrow).

677. Cervical rib, operative finding. It is lying behind the subclavian artery.

678. Arteriogram showing compression of subclavian artery on abduction of the arm. This narrows the space between the clavicle and the first rib, i.e. costoclavicular syndrome.

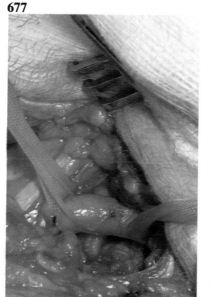

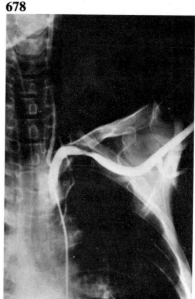

679 & 680. Shoulder/hand syndrome. Sixteen weeks after a myocardial infarct this man developed pain, swelling, and stiffness in his left hand. The fingers later became smooth, dusky, and pink. His initial complaint was of pain in the shoulder. In this condition the palmar fascia often thickens and x-rays may show osteoporosis.

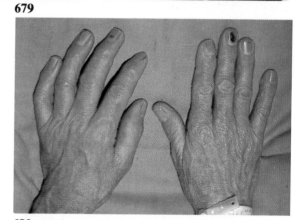

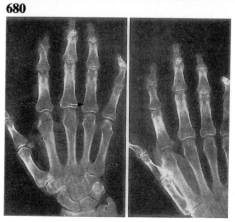

Acrocyanosis. A chronic vasospastic disorder restricted to females and characterised by persistent cyanosis of the hands and feet. In severe conditions there is pain and numbness and occasionally hyperhidrosis. Peripheral pulses diminish in the cold but return with re-warming.

681. **Acrocyanosis** of the left hand. Recent carpal tunnel decompression on the right hand.

681

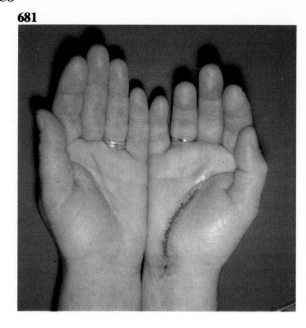

Veins and lymphatics

682. The dorsal veins of the hand.

(1) basilic vein
(2) dorsal venous arch
(3) superficial dorsal veins
(4) intercapitular veins
(5) dorsal digital veins
(6) perforating vein
(7) cephalic vein

683. Superficial thrombophlebitis caused by an indwelling catheter on the dorsum of the right hand. The left hand had been shaved for surgery.

683

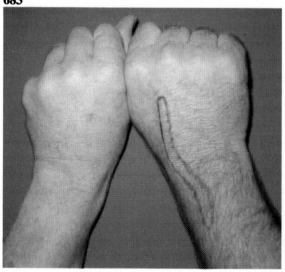

682

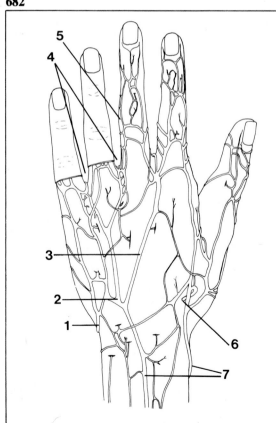

684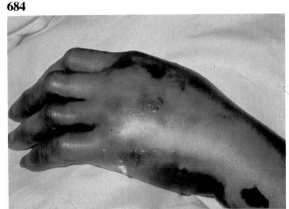

684. Major venous occlusion. Axillary vein thrombosis (effort thrombosis, Paget-Schroetter's syndrome) in a 22-year-old Negro. Rapid swelling of the arm and hand with pain, cyanosis, coolness, and muscular weakness. The arterial circulation was unaffected. Normal function returned with conservative treatment.

685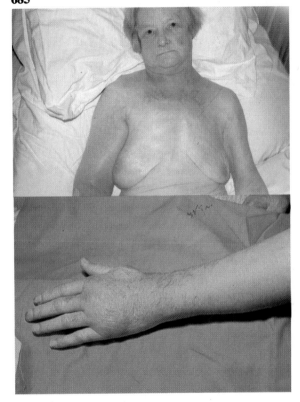

685. Superior vena cava thrombosis. Venous congestion of the hand and arm.

686. Varicose veins of the right middle finger and to a lesser extent the other fingers.

687. Congenital arteriovenous fistula in the right ring finger. It is commonly associated with haemangioma and increases the circulation to the bony epiphyses to produce increased growth. There may be enlarged veins, oedema, sweating, and ulceration.

An arteriogram shows simultaneous filling of the arteries and veins and prominent vascular changes in the cortex.

686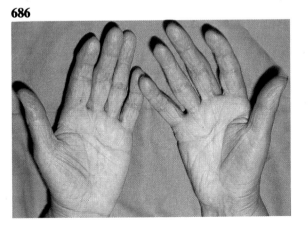

687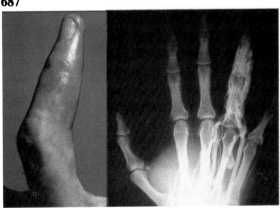

Oedema

Oedema is most notable on the dorsum where the tissues are loose and lax.

Bilateral oedema of the hand occurs in cardiac, renal and hepatic failure, superior vena caval obstruction, or in hypoproteinaemia.

Unilateral oedema of the hand may be caused by obstruction of the venous or lymphatic return. It can also occur after injury, infection or allergy.

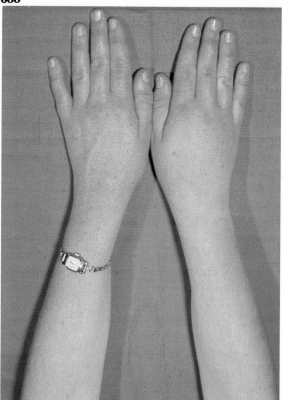

688. Primary lymphoedema on the right hand of a 23-year-old woman. Similar lymphoedema can occur after inflammatory blockage or surgical resection of the draining lymph nodes. Progressive stasis of high protein fluid results in secondary fibrosis, and episodes of acute and chronic inflammation can be superimposed with further stasis and fibrosis. Eventually hypertrophy of the limb results, with marked thickening and fibrosis of the skin and subcutaneous tissues and diminution.

Venous oedema from superior vena cava obstruction (see **685**).

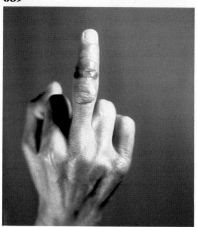

689. Artefact oedema. Self-inflicted oedema of the right middle finger produced by applying a tight band around the finger (see also **356–358**).

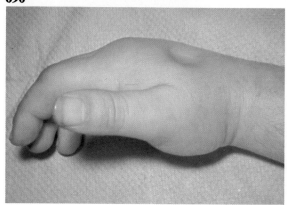

690. Dorsal oedema in the case of palmar hand infection (see **372, 391**).

691. Post-operative oedema. Twenty-four hours after operation on the carpometacarpal joint of the thumb, this 45-year-old woman developed this significant amount of oedema. She had not had adequate elevation after operation.

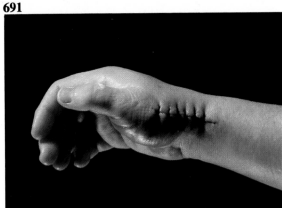

692. Acute post-traumatic oedema. There is contusion of the skin and oedema of the right forearm, wrist and hand.

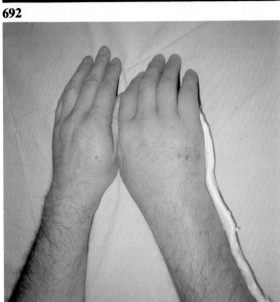

693. Chronic post-traumatic oedema. After injury oedema may form on the dorsum of the hand, but rather than gradually fading away proximally, there is a ridge at the wrist representing the upper extremity of the oedema. Most of these patients complain of pain and disability, but they usually have a full range of movement without muscle weakness or wasting. Similar oedema can be found in hysterical patients.

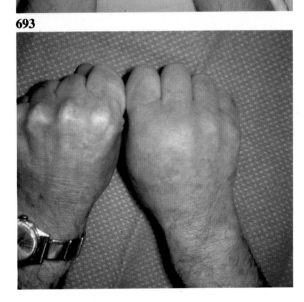

Angioneurotic oedema. May present without apparent cause, although allergy and psychological factors may play a role. The oedema is most marked on waking and disappears as the day goes on. Pitting is easily produced.

8. Painful disorders

Definition of pain. Pain is a disagreeable sensation mediated by a complex system of afferent nerve stimuli which interact with the emotional state of the individual. Pain is modified by a person's past experience, motivation, and state of mind.

The hand is one of the body's most sensitive structures, sensation being one of its major functions. Pain in the hand usually reflects a disorder of one of its gliding parts such as a tendon or joint, an irritation of a peripheral nerve, a lack of blood supply to the hand, or infection. See Chapter 4.

Pain is the guardian of the hand against injury. Congenital absence of pain makes the hand susceptible to the stresses and injuries of everyday life, and the lack of the pain protective mechanism results in loss of blood supply, infection, gangrene, and amputation of those parts subjected to stress.

694. Congenital absence of pain in a 15-year-old boy. Repeated injury and infection had resulted in loss of his fingertips, right foot, and the toes of his left foot.

695 & 696. Branches of the brachial plexus which innervate the hand (see **26**). A complete examination may necessitate an investigation into the functions of the spinal cord, cervical vertebral column, and the pathways of the brachial plexus from the neck to the hand.

694

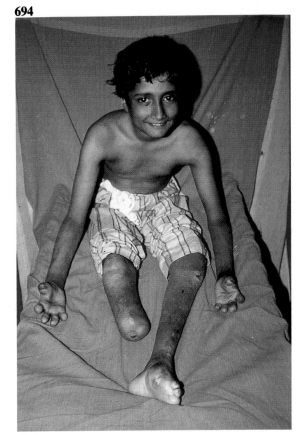

696

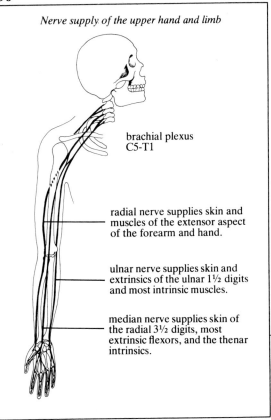

Nerve supply of the upper hand and limb

brachial plexus
C5-T1

radial nerve supplies skin and muscles of the extensor aspect of the forearm and hand.

ulnar nerve supplies skin and extrinsics of the ulnar 1½ digits and most intrinsic muscles.

median nerve supplies skin of the radial 3½ digits, most extrinsic flexors, and the thenar intrinsics.

695

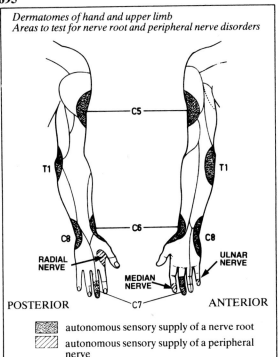

Dermatomes of hand and upper limb
Areas to test for nerve root and peripheral nerve disorders

C5

T1   T1

C6

C8   C8

RADIAL
NERVE   ULNAR
                      NERVE

MEDIAN
NERVE

POSTERIOR   C7   ANTERIOR

▨ autonomous sensory supply of a nerve root
▨ autonomous sensory supply of a peripheral nerve

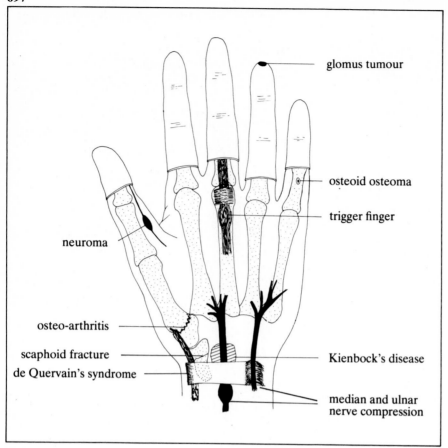

glomus tumour

osteoid osteoma

trigger finger

neuroma

osteo-arthritis

scaphoid fracture

de Quervain's syndrome

Kienbock's disease

median and ulnar
nerve compression

697. Differential diagnosis of pain in the hand.

History

(a) Site and radiation. Pain felt in the hand and radiating proximally up the limb usually arises from the hand itself. For example, in the carpal tunnel syndrome pain may radiate to the neck, but there are no physical signs proximal to the wrist. Pain originating in the neck, for example cervical spondylitis, shoots down the limb.

(b) Type of pain. A throbbing pain usually indicates venous or lymphatic congestion, the commonest being inflammatory oedema or external compression from a tight bandage or plaster. A local burning pain usually indicates the pain of a pressure sore.

Persistent generalised burning pain is found in causalgia. The pain of nerve compression may also be described as burning, but there is a strong 'paraesthetic' element and this pain is intermittent. A neuroma causes a sharp shooting pain on contact.

(c) Onset and aggravation. Pain and numbness waking the patient from sleep is characteristic of a carpal tunnel syndrome. Movement aggravates the pain of any inflamed tissue.

(d) Relief of pain. Rest in a splint relieves the pain of most conditions. Elevation alleviates the pain of congestion. Salicylates are said to alleviate the pain of an osteoid osteoma.

(e) Pattern of pain. An intermittent pain is characteristic of a carpal tunnel syndrome, corresponding with periods of fluid retention. Causalgia produces a continuous pain.

Examination

(a) Look for a scar. This may be the clue to an underlying neuroma (palmar branch of median nerve, amputation neuroma).

Look for muscle wasting. Abductor pollicis brevis in carpal tunnel syndrome; brachio radialis in cervical spondylitis.

Look for vasomotor changes. Causalgia; Raynaud's syndrome.

Look for postural deformity. Subluxation of the carpometacarpal joint of the thumb.

Look for signs of infection. (Chapter 4).

(b) Feel for crepitus. Tenosynovitis, osteoarthritis, non-united fracture.

Feel for local tenderness. Tinel's sign over a neuroma.

Investigations

(a) X-ray for: bone and joint change – arthritis; tumour; osteoid osteoma, **556**; acute calcific tendinitis, **449**; infection.

(b) Other tests: serum uric acid – gout.

Differential diagnosis of pain in the hand – sites and causes

---
THE HAND ITSELF

Skin	Pressure sore
Subcutaneous	Infection
Fascia	Fasciitis
Tendon	*Tendon constriction*, tenosynovitis, tendinitis
Nerve	Nerve compression (*carpal tunnel syndrome*), neuroma, peripheral neuritis, reflex dystrophy
Blood vessels	Ischaemia, congestion, glomus tumour
Joint	Synovitis, gout, *arthritis*
Bone	Osteomyelitis, osteoid osteoma, Kienbock's disease Sudeck's atrophy, fracture

---
REFERRED FROM

Neck	*Cervical spondylosis* etc
Thoracic viscera	Ischaemic heart disease
Shoulder, elbow	Shoulder/hand syndrome

MISCELLANEOUS

	Functional neurosis Hysteria

Diagnosis of pain in the hand – history

Site. Finger, hand, wrist or arm?
Radiation. Proximal or distal?
Type. Throb, burn or paraesthesia?
Onset. Night, morning, or on exercise?
Relief. Splint, elevation or analgesics?
Pattern. Continuous or intermittent?

Pain arising in the arm and hand

Skin

The skin with its many pain receptors in the dermis provides an elastic, protective, sensitive cover for the vital structures of the hand. Direct injury to the skin or secondary damage from ischaemia will cause pain. A common cause of pain is a pressure sore from a tight dressing or splint over unpadded bony points.

698. Pressure sore. This patient complained of persistent burning pain beneath the bandage.

699. Pressure sore. Findings after removal of the bandage. A pressure sore over the lower end of ulna.

700. A pressure sore from the rubbing of adjacent proximal interphalangeal joint. Persistent pain.

701. Pressure sores from an unpadded splint.

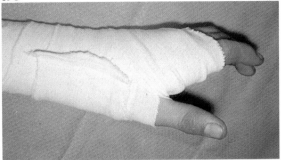

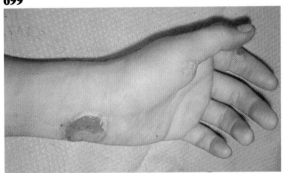

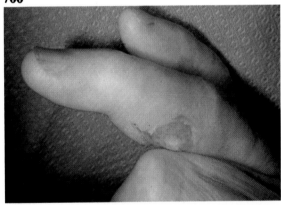

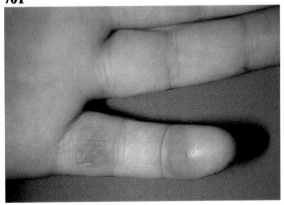

Fascia

The palmar fascia has its own pain receptors and when stretched it becomes painful.

Painful palmar fasciitis can occur in Dupuytren's contracture. It also occurs after injury, operation, coronary occlusion, 'stroke', and in the shoulder/hand syndrome.

702. **Painful palmar fasciitis** with increasing fibrosis and nodule formation in a 40-year-old journalist who has Dupuytren's diathesis, see page 280.

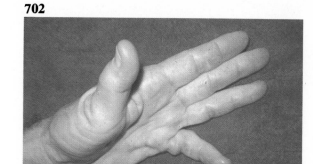

703. **Painful palmar fasciitis** in a 70-year-old, three months after a coronary occlusion.

704. **Palmar fascial bands**, which are prominent in both hands are seen here in the left.

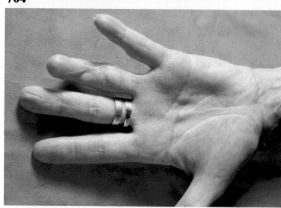

705. **Painful palmar fasciitis** (arrow) developed in this 32-year-old woman after operation to release a trigger finger.

706. **Painful palmar fasciitis** with prominent palmar fascial bands in a 22-year-old woman, developing three months after laceration of her palm. Because of pain she kept her fingers cupped.

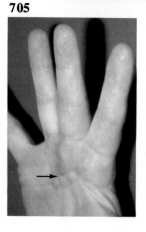
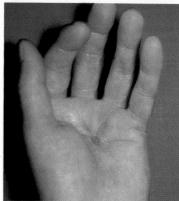

Secrétan's disease. This is a dorsal peritendinous fibrosis which occasionally develops after a relatively minor contusion to the dorsum of the hand. It is associated with dorsal swelling, limitation of movement, and an aching discomfort in the hand.

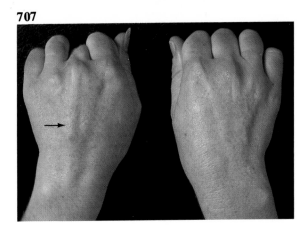

707. Secrétan's disease over the left middle metacarpal in a 26-year-old woman. Stiffness of the middle finger.

Henri François Secrétan (1856–1916). Swiss medical specialist in insurance and accident work. In 1901 he described what he called hard oedema and traumatic hyperplasia of the dorsum of the metacarpus following a single violent contusion.

Tendon

Muscle tendon units can become painful when they are constricted, contracted, stretched or when they are subjected to repeated friction.

Tendon constriction syndromes. Stenosing tenosynovitis. Both flexor and extensor tendons pass through fibrous and synovial tunnels at the wrist, and flexor tendons pass through similar tunnels along their course in the fingers. Disproportion between the tendons and their tunnels is caused by:

(1) Enlargement of the tendon or its synovial sheath (as in frictional or rheumatoid synovitis), or

(2) Constriction of the fibrous tunnel which can be developmental or post-traumatic.

In all instances oedema of both tendons and tunnels plays a major role. For this reason these syndromes occur in middle-aged or elderly women, either after rest or sleep or after periods of excessive, vigorous, resisted effort.

Clinically there is pain, crepitus, triggering or locking. The most common types of stenosing tenosynovitis are trigger finger or thumb, and de Quervain's syndrome.

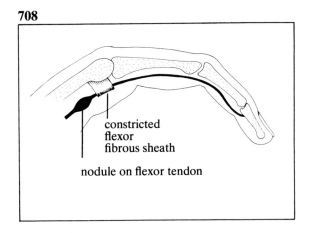

708. Trigger finger or thumb. Diagram of trigger digit. This occurs either from a thick tendon or a constricted pulley at the metacarpo-phalangeal joint. The powerful flexor muscles can pull the thickened tendon through the constricted pulley into the palm. Depending on the disproportion, the relatively weak extensor mechanism may reverse the direction with a painful click, such pain radiating to the proximal interphalangeal joint, or the finger may stay locked in flexion.

When the disproportion prevents all tendon glide, the position of the swelling of the tendon in relation to the pulley determines the position of the finger. If it lies proximal to the pulley, the digit will be flexed and cannot be extended. If the swelling is distal to the pulley the digit can be passively but not actively flexed.

709. Trigger finger in an adult. (Usually the ring or middle finger.) Tender lump at the base of the finger over the metacarpophalangeal joint (arrow). The patient is unable to actively extend the finger. Passive extension produces pain radiating to the proximal interphalangeal joint.

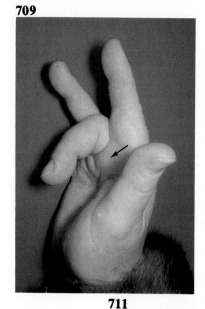

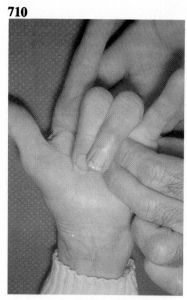

710. Testing for trigger finger. Feel for a lump and crepitus as the patient flexes and extends the finger.

711. Rheumatoid flexor tenosynovitis. An example of flexor tendon enlargement causing painful crepitus and triggering. See also **473**.

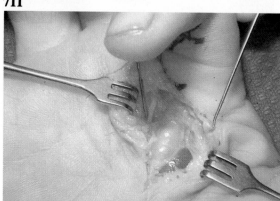

712 & 713. Trigger thumb. Before and after release of the constricted fibrous sheath of flexor pollicis longus in a professional fisherman.

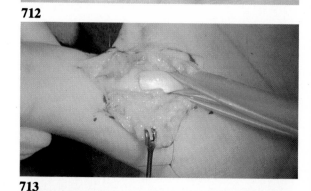

de Quervain's syndrome (chronic stenosing teno-synovitis).

714. de Quervain's syndrome. There is constriction of extensor pollicis brevis and abductor pollicis longus in the first extensor compartment over the radial side of the distal end of the radius. It is the tendon sheath which is thickened and not the tendons themselves. There is pain, local swelling, and sometimes irritation from oedema of the over-lying superficial sensory branches of the radial nerve. It occurs most often in middle-aged women after excessive repetitive movements of the thumb as in sewing, wringing of wet clothes etc.

There are four signs:

(1) Localised tenderness over the radial side of the lower end of the radius, extending 1.5cm above the tip of the styloid process.

(2) Pain on active abduction and extension of the thumb.

(3) Passive extension of the thumb is painless.

(4) 'Finkelstein's test'. Ask the patient to flex his fingers over his flexed and adducted thumb. Ulnar deviation of the wrist produces pain over the radial side of the wrist.

714

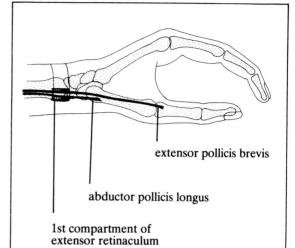

extensor pollicis brevis

abductor pollicis longus

1st compartment of extensor retinaculum

Note:

Pain arises from the constricted tendons or from irritation of the overlying branches of the radial nerve.

There are often two or more tendons of abductor pollicis longus.

Extensor pollicis brevis may be in a separate compartment to abductor pollicis longus.

715. de Quervain's syndrome. The arrow points to the area of local and swelling tenderness.

715

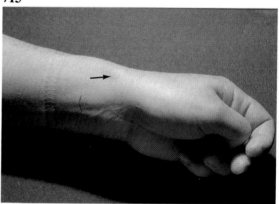

716. Both wrists compared. Thickened tendon sheath on the left side. Frictional or occupational tenosynovitis. (See Chapter 9).

716

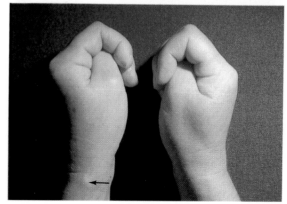

Fritz de Quervain (1868–1940). Swiss general surgeon who described his syndrome, i.e. occupational chronic tenovaginitis, in 1895 whilst working in the watch-making district of La Chaux-de-Fonds.

717. Acute calcification or calcific tendinitis. Calcification in the hypothenar muscles. Acute calcification can occur in the tendons or ligaments of the wrist, hand or fingers, usually in patients 30 to 60 years old. They present with sudden severe pain and inflammation resembling acute cellulitis or arthritis. The lymph glands are not involved. X-ray shows a calcium deposit.

717

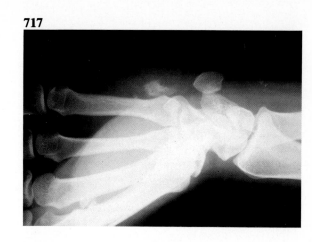

Nerves

Neurogenic pain in the hand is mostly from peripheral nerve injury, irritation or compression, and less often from peripheral neuritis or central nervous system disorders.

For differential diagnosis between nerve root and more peripheral nerve disorders see the Appendix, pages 338, 339.

COMPRESSION SYNDROMES

718. The nerve roots destined for the hand and the three major nerves of the hand itself, namely the median, ulnar, and radial nerves, can be compressed within anatomical tunnels in their course from the neck through the limb.

The causes are:

(1) Reduction in size of the tunnel, e.g. congenital or traumatic deformity – fracture, dislocation or scar.

(2) Increase in the contents of the tunnel by oedema, tumour or proliferative tenosynovitis.

The pathological condition of the nerve is initially that of neuropraxia, but this may become intractable axonotmesis. The clinical presentation will depend on whether the compression is acute, chronic, intermittent or progressive.

In a mixed nerve such as the median nerve there will be three types of clinical signs:

(1) Sensory – pain and paraesthesia.

(2) Motor – weakness followed by palsy.

(3) Autonomic – vasomotor changes. Vasodilatation or constriction. Disorders of sweating.

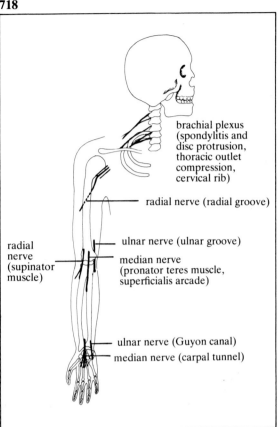

brachial plexus (spondylitis and disc protrusion, thoracic outlet compression, cervical rib)

radial nerve (radial groove)

ulnar nerve (ulnar groove)

radial nerve (supinator muscle)

median nerve (pronator teres muscle, superficialis arcade)

ulnar nerve (Guyon canal)

median nerve (carpal tunnel)

719

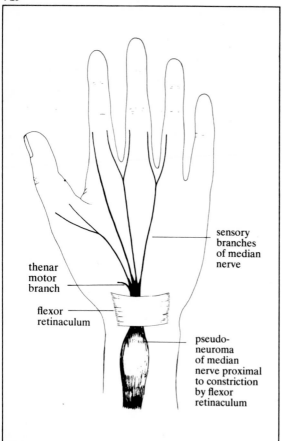

thenar
motor
branch

flexor
retinaculum

sensory
branches
of median
nerve

pseudo-
neuroma
of median
nerve proximal
to constriction
by flexor
retinaculum

Median nerve compression. This most commonly occurs beneath the flexor retinaculum, as in the carpal tunnel syndrome. It can also be compressed in the forearm by the pronator teres or flexor superficialis muscle arcade.

719. Carpal tunnel and carpal tunnel syndrome. See also **822**. Carpal tunnel syndrome is a compression neuropathy of the median nerve as it passes beneath the flexor retinaculum at the wrist. It occurs mostly in women in the 5th or 6th decades and in 50% of cases the symptoms are bilateral, although they are more evident in the dominant hand. The patient complains of pain and paraesthesia in the median nerve distribution of the hand, but this pain can radiate up the limb to the shoulder or neck. Abnormal physical signs should not occur proximal to the wrist. Characteristically, the pain wakes the patient from sleep. This is called 'waking numbness'. The patient also complains of progressive weakness or clumsiness, due to impairment of the finer movements of the hands from thenar muscle branch compression.

Carpal tunnel syndrome. Compression median neuropathy, first described by Sir James Paget in 1853.

720

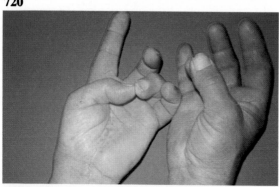

720 & 721. Clinical features of carpal tunnel syndrome. For six years this 45-year-old labourer had painful paraesthesia in the left hand. Note the lack of work staining of the skin of the left thumb, index and middle fingers, and also the thenar muscle wasting and the lack of opposition.

722. Operative findings. Compression of the median nerve.

721

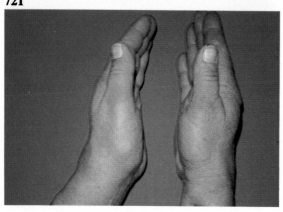

722

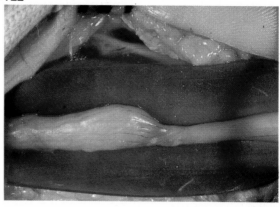

723. Tinel's sign at the wrist. Percuss along the course of the median nerve. When a pseudo-neuroma is present from carpal tunnel compression, percussion over it will produce a strong sensation of pins-and-needles which radiates to the fingers.

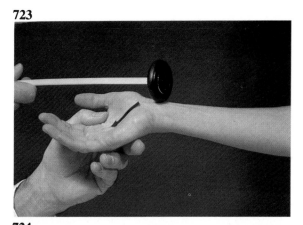

724. Phalen's wrist flexion test. Holding the wrists in forced flexion may cause the flexor retinaculum to impinge on the median nerve and aggravate the paraesthesia in the affected fingers.

The tests should be timed, the more rapid the onset of paraesthesia the more definite is the diagnosis.

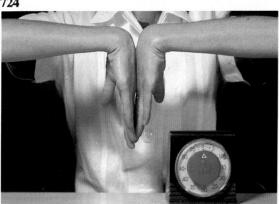

Pronator syndrome. The median nerve can be compressed by the pronator teres, the ridge of origin of flexor superficialis or by anterior dislocation of the head of the radius. Clinical features are similar to those of the carpal tunnel syndrome but there will be an absence of the Tinel and Phalen signs at the wrist. Instead there will be a Tinel sign over the median nerve at the junction of the proximal and mid two-thirds of the forearm.

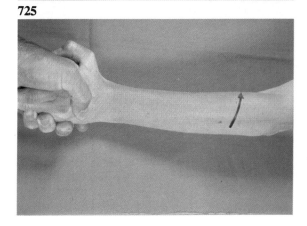

725. Pronator syndrome. Pain and tingling may be produced in the pronator syndrome when the patient attempts to pronate the extended forearm against resistance.

Anterior interosseous syndrome. The anterior interosseous nerve arises from the median nerve about 4–6cm below the elbow. Although there are no sensory branches compression of this nerve near its origin can produce pain in the forearm. Diagnosis is made by finding weakness of flexor pollicis longus and flexor digitorum profundus to the index finger.

Jules Tinel (1879–1952). French physician. Described a test for regenerating sensory axons. The test does not become positive until six to eight weeks after a wound or repair of a peripheral nerve. Then, percussion over that site produces a tingling pain.

Ulnar nerve compression. The ulnar nerve can be compressed at two places, at the elbow and at the wrist. At the elbow the nerve can be compressed in the ulnar groove behind the medial epicondyle or in the cubital tunnel as the nerve passes through the aponeurosis of flexor carpi ulnaris.

Entrapment may be caused by:

(1) pressure or friction,

(2) diminished size of the tunnel as in osteo-arthritis, fracture or skeletal deformity, or

(3) an increase in content of the canal from oedema or a space occupying lesion, e.g. a ganglion.

The clinical features include:

(1) Numbness, dysthesia and pain passing down the forearm into the ulnar distribution of the hand.

(2) Loss of dexterity, grip and pincer strength.

The motor signs will depend on the level of compression. Compression behind the medial epicondyle will affect the flexor carpi ulnaris and the profundus tendon to the ring and little fingers, as well as the 15 ulnar-innervated intrinsic muscles. Compression in the cubital tunnel will not affect the flexor carpi ulnaris.

Ulnar groove compression of the ulnar nerve, i.e. 'tardy ulnar palsy' – so called because the ulnar palsy often follows some years after an injury which may have produced a cubitus valgus deformity of the elbow with deformity of the ulnar groove.

726 & 727. Cubitus valgus and ulnar nerve compression in a 53-year-old woman who fractured her elbow as a child.

Occupations requiring forced repetitive flexion and extension of the elbow may cause frictional neuritis of the ulnar nerve.

728. Ulnar neuritis at the elbow. When the ulnar nerve is palpated in the groove or in the cubital tunnel the patient often experiences paraesthesia radiating down to the little finger. There may be a positive Tinel sign.

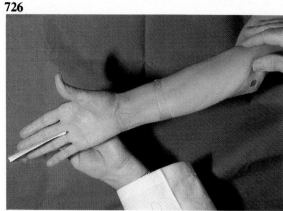

726

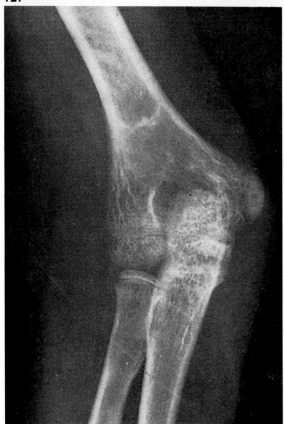

727

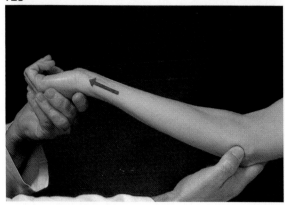

728

729. Forceful flexion of the elbow in ulnar neuritis at the elbow will also cause paraesthesia.

Compression of the ulnar never at the wrist may occur from a ganglion. Mostly there is compression of the deep motor branch with weakness of grasp and pinch. Complete ulnar nerve involvement with sensory as well as motor signs is uncommon.

729

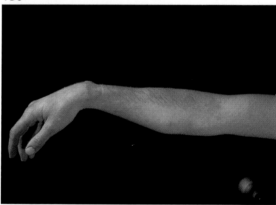

730. Compression of the radial nerve. Saturday night palsy. Sleeping with an arm resting over the back of a chair may result in compression of the radial nerve against the humerus in the spiral groove. The patient presents with a wrist drop from paralysis of the wrist and finger extensors. There is also numbness over the radial side of the back of the hand.

730

Compression of the dorsal interosseous nerve. 'Radial tunnel syndrome'. The posterior interosseous nerve is sometimes compressed in its course through the supinator muscle. The patient complains of pain, usually localised to the extensor mass below the elbow and thus mimicking 'tennis elbow' or lateral epicondylitis. The pain may radiate to the dorsal aspect of the wrist. Sensory and motor signs are rare but there is ordinarily tenderness along the course of the posterior interosseous nerve.

731. Radial tunnel syndrome. Test for tenderness along the posterior interosseous nerve.

Resisted supination may also produce pain radiating along the post-interosseous nerve to the wrist.

731

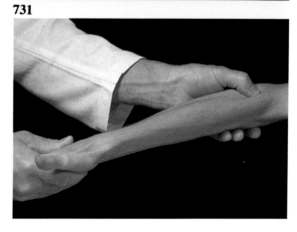

Reflex dystrophies

732. Reflex dystrophies of the hand. Reflex dystrophies are of two types:

(1) organic–causalgia, Sudeck's atrophy, shoulder/hand syndrome, and

(2) psychogenic.

They are all associated with pain and show various vasomotor changes from vasodilatation to vasoconstriction. They result finally in disuse, with trophism and osteoporosis.

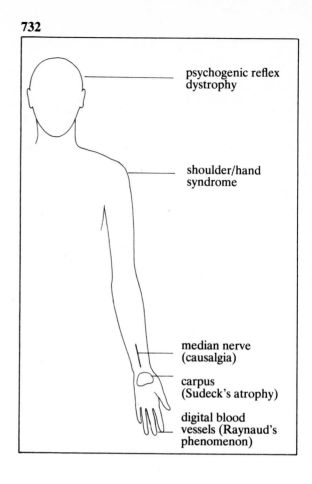

psychogenic reflex dystrophy

shoulder/hand syndrome

median nerve (causalgia)

carpus (Sudeck's atrophy)

digital blood vessels (Raynaud's phenomenon)

Types and clinical features of reflex dystrophies

CONDITION	CAUSE AND ONSET	CLINICAL AND X-RAY FEATURES
Causalgia	Days or weeks after injury to the hand or limb Can follow median nerve injury	Constant burning pain evoked by: light touch, emotion, temperature, cold, noise, and vibrations *Four stages:* (1) vasodilatation – red, smooth *Then* (2) vasoconstriction – blue *Leading to* (3) osteoporosis *Finally* (4) ischaemic contracture *Findings:* hyper-irritation and hyper-aesthesia in the distribution of the median nerve. Trigger areas *General health* may be affected *X-ray:* osteoporosis *Relief* by sympatheic ganglion blockade
Sudeck's atrophy	Sprain or fracture of the wrist	Dull ache or pain Vasomotor changes occur as in causalgia; i.e. vasodilatation, then vasoconstriction *Findings:* puffy, patchy, perspiring, hyper-aesthetic hand; palmar fasciitis *General health* not affected *X-ray:* patchy rarefaction of carpus *Relief* by sympathetic ganglion blockade
Shoulder/hand syndrome	Injury to the shoulder or upper limb Prolonged immobilisation of the shoulder	Painful stiff shoulder which becomes 'frozen' Painful, swollen, stiff hand with palmar fasciitis Vasomotor changes occur as in causalgia; i.e. vasodilatation then vasoconstriction Relief by sympathetic ganglion blockade
Psychogenic reflex dystrophy *(This is a dangerous primary diagnosis)*	Usually the basis is an underlying emotional conflict e.g. fear of loss of job or hope of financial gain; often follows minor injury	Sensory symptoms less often associated with one peripheral nerve. General health rarely affected Poor response to sympathetic ganglion blockade

733

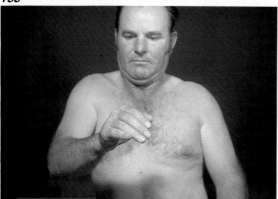

734

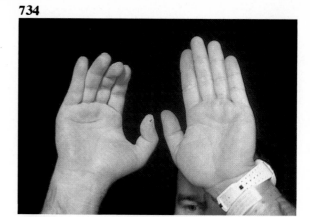

735

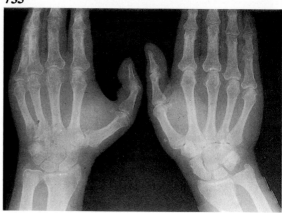

733–735. Causalgia. Causalgia in a 45-year-old forester who one year before had a tree fall on his right upper limb. The right median nerve was severely contused at the wrist. He developed a persistent, severe, continuous burning pain extending from his fingertips to his shoulder and neck.

His hand (**734**), too painful to use, shows wasting, increased sweating and a flexion contracture of the thumb. The x-ray shows osteoporosis of the carpus.

Paul Hermann Martin Sudeck (1866–1938). German surgeon. Sudeck's atrophy, Kienboeck's atrophy, reflex sympathetic dystrophy, reflex bone atrophy. Osteoporosis, usually of the wrists, hands, and feet in association with swelling and tenderness of underlying soft tissues, which follows fractures or minor injuries. Autonomic vasomotor disorders are the suspected cause.

736 & 737. Sudeck's atrophy. Two weeks after a fall on his left wrist this man developed a puffy, painful, stiff, sweaty and hypersensitive left wrist and hand. The x-ray shows patchy rarefaction of the carpal bones.

736

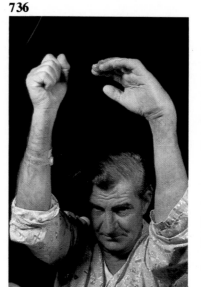

737

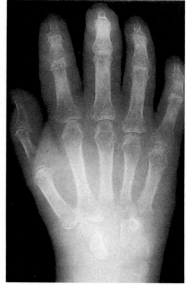

Shoulder/hand syndrome. A reflex dystrophy in which both shoulder and hand become painful and stiff. Vasomotor changes occur in the hand, first vasodilation and later vaso-constriction. The end result is often a hand that is swollen with shiny trophic skin and with demineralisation of the bone. Palmar fascial contractures may develop.

738. Shoulder/hand syndrome in a 70-year-old man, three months after myocardial infarction.

739. Spotty osteoporosis in the hand and wrist. The x-ray is non-specific but it may help in the diagnosis when the clinical signs are present.

738

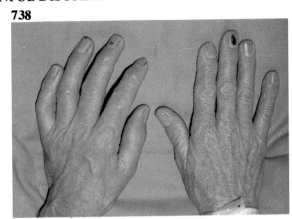

739

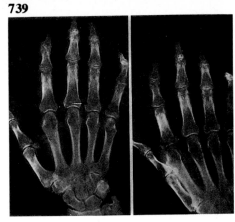

Neuroma

740. A neuroma. A terminal neuroma or neuroma in continuity will present clinically as painful paraesthesia or dysthesia when that part of the hand or finger containing them is knocked.
See Tinel's sign, **723**.

740

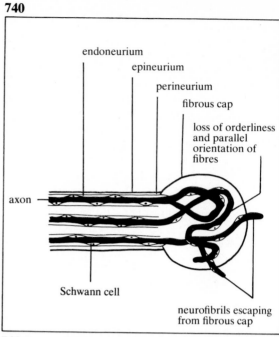

741. Neuroma in an amputation stump. Electric shock feelings when the stump is knocked.

741

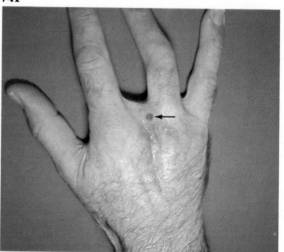

742. Operative findings.

742

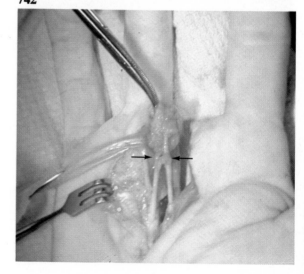

743. Neuroma in continuity of the palmar branch of the median nerve from injury. This patient was unable to use her wrist or hand because of pain on contact with the scar over the neuroma.

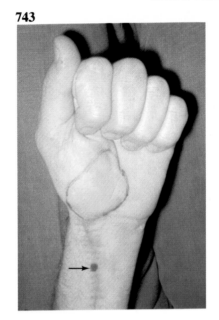

Peripheral neuritis. (See Appendix, pages 338, 339.) Peripheral neuritis is most common in alcoholics and diabetics. The patient complains of pain, paraesthesia or numbness which may begin in the hands and feet but then spreads to the trunk. Motor weakness and vasomotor changes follow. The deep reflexes, originally increased, are now lost.

The diagnosis of peripheral neuritis will be suggested if the pains in the limbs are associated with marked sensory changes – anaesthesia, paraesthesia, and hyperaesthesia – with tenderness of the skin and muscles, or along the course of the nerves, and with weakness, atrophy, and diminished or absent reflexes.

Herpes zoster (shingles). This is an acute viral infection of nerve structures producing groups of epidermal vesicles distributed along the course of one or more peripheral nerves on one side of the body. The inflammatory changes are usually confined to one posterior root ganglion.

Pain of varying degrees of severity in the distribution of the nerve root precedes the vesicular eruption by a day or two.

744. Herpes zoster in a 45-year-old patient who presented with pain.

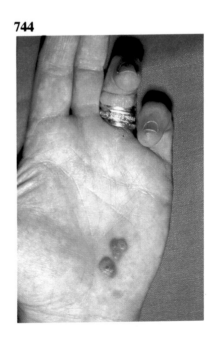

Blood vessels

Vascular causes of pain in the hand include arterial insufficiency, venous congestion and inflammation. The glomus tumour is also a cause of pain.

745. **Painful digital ischaemia** in the left index and middle fingers of an 18-year-old female with scleroderma. Slow healing sores are seen at each fingertip. She has persistent pain at rest.

746. **Acute ischaemia** causing severe pain in the little finger. Ulnar artery embolus.

747. **Glomus tumour,** presenting as persistent severe pain at the fingertip of a 30-year-old house-wife. Apart from local tenderness and disuse atrophy of the pulp, there were no abnormal physical signs. She was diagnosed as having hysteria. She avoided using the pulp and cutting the fingernail. The operative finding was a glomus tumour associated with a digital neurovascular bundle.

748. **Painful congestion** of the wrist and forearm after crush injury.

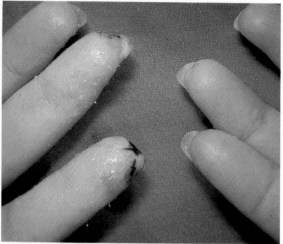

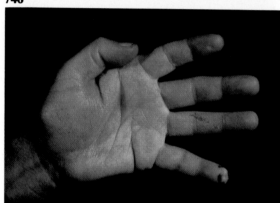

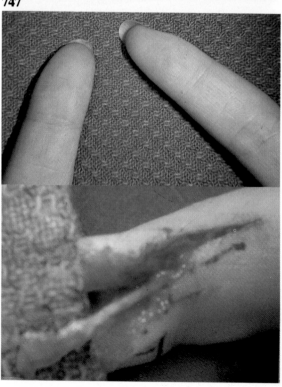

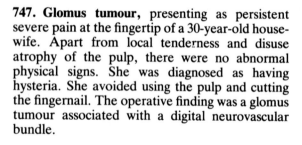

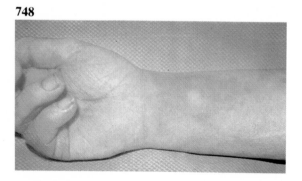

Joints

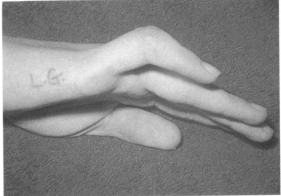

749. Acute synovitis. Following injury, the proximal interphalangeal joint and to a lesser extent the metacarpo-phalangeal joint develop acute synovitis characterised by pain, swelling and joint stiffness. The proximal interphalangeal joint stiffens in flexion, the metacarpo-phalangeal joint stiffens in extension.

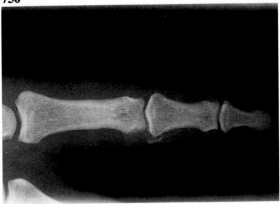

750. Acute peri-articular calcification following injury, presenting as acute on chronic arthritis.

Gout. (See Chapter 10). Acute gout is characterised by severe pain with a purple red inflammatory area resembling cellulitis. It can affect any joint, bursa or tendon sheath. Its onset may be associated with alcoholic intake. In 75% there is a raised uric acid, and in 50% the condition begins with podagra.

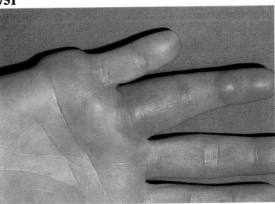

751. Gout affecting the soft tissues of the left ring finger.

752. Gouty arthritis and synovitis with secondary carpal tunnel syndrome.

753. Operative findings. Urate crystals on the capsule of the wrist joint.

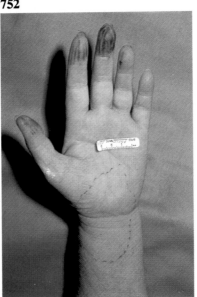

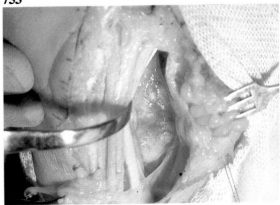

Osteo-arthritis. (See Chapter 10.) Osteo-arthritis is a degenerative arthritis occurring in the older age groups, from wear and tear of the cartilage, with secondary reactionary formation of osteophytes from the underlying bone. It occurs especially in the distal interphalangeal joints and in the carpo-metaphalangeal joint of the thumb. The patients complain of pain and stiffness, worse after sleep and rest. The physical signs include diminished movement and crepitus.

754. Osteo-arthritis in the distal interphalangeal joints of a 55-year-old woman.

755. Osteo-arthritis, x-ray.

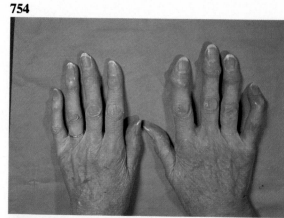

754

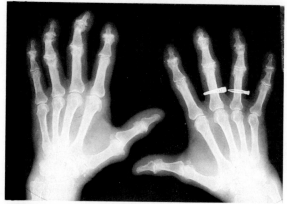

755

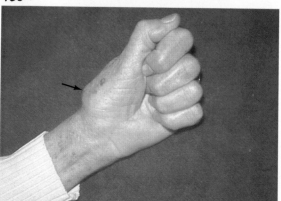

756

756. Osteo-arthritis in the carpo-metacarpal joint of the thumb in a leather worker.

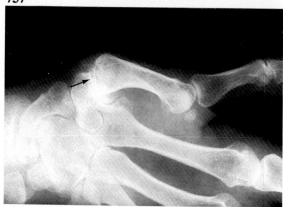

757

757. Narrowed joint spaces, irregular joint margin and osteophyte formation.

Rheumatoid arthritis. (See Chapter 10.) This is initially a rheumatoid synovitis affecting the more proximal finger joints, and the metacarpophalangeal and proximal interphalangeal joints in the younger age groups. It also affects the wrist. Inflammatory granulation tissue erodes the capsule and subchondral bone and distends the ligaments and capsule of the joint. The patients complain of pain as in acute synovitis. Stiffness and deformity follow in the chronic stage.

758. Rheumatoid synovitis presenting as pain and swelling in the proximal, interphalangeal and metacarpo-phalangeal joints.

759. Rheumatoid synovitis. There are no abnormal x-ray features at this early stage.

758

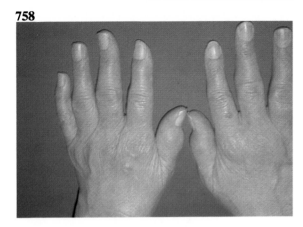

759

Bone

760 & 761. Osteoid osteoma presenting as pain and later as a swelling in the right ring fingertip. The x-ray shows a small nidus of rarefaction. The central sclerotic focus is marked by an arrow.

760

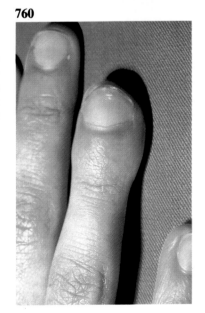

761

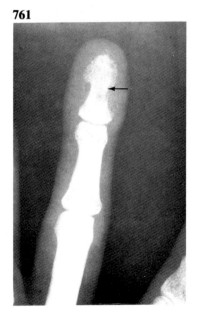

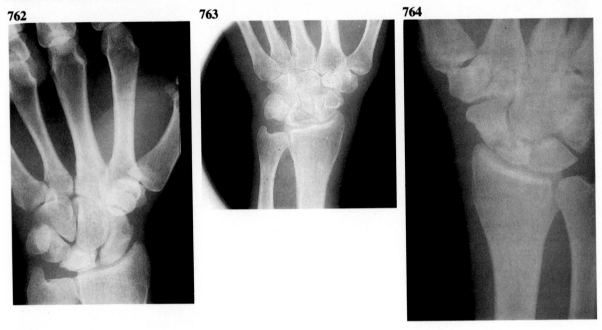

762   **763**   **764**

Kienboeck's disease. Aseptic necrosis of the lunate can occur in young adults long after a trivial or repeated injury to the wrist. Radiologically there may be osteosclerosis or osteoporosis.

762. Kienboeck's disease. X-ray showing sclerosis of the lunate and associated osteo-arthritis at the radiocarpal joint. This patient presented with persistent aching discomfort, with severe pain on vigorous exercise.

763. Kienboeck's disease. X-ray showing osteoporosis of the lunate.

764. Aseptic necrosis of the scaphoid after fracture. The blood supply to the proximal pole is cut off.

Sudeck's atrophy. Weeks or months after a trivial injury to the wrist or distal forearm the hand and fingers become painful, puffy, stiff, patchily discoloured, sweaty and hypersensitive. An x-ray shows a patchy rarefaction of the carpal bones probably due to disuse. See **669–671**.

Robert Kienboeck (1871–1953). Austrian radiologist who used the newly discovered x-rays to establish diagnostic and therapeutic radiology as a special field of study. Described lunate malacia, Kienboeck's disease, in 1910.

Referred pain

765. Pain may be referred to the hand from the neck or thorax.

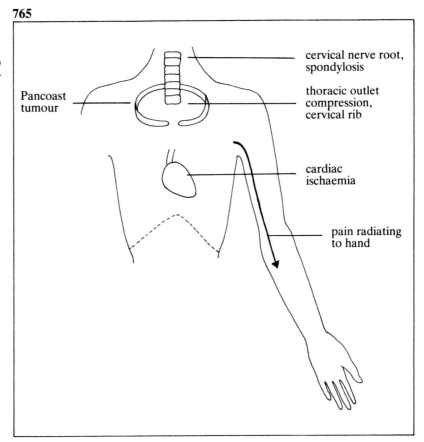

Pancoast tumour

cervical nerve root, spondylosis

thoracic outlet compression, cervical rib

cardiac ischaemia

pain radiating to hand

Referred pain from the neck

Brachial neuralgia, i.e. pain and tenderness in the distribution of the brachial plexus without paralysis of muscles or sensory loss, and without any gross lesion being found to account for the symptoms. Can occur in rheumatism, gout, alcoholism, diabetes or influenza.

It differs from muscular rheumatism or fibrositis where the pain and tenderness are referred more to the insertion of muscles.

Brachial neuritis. Muscular atrophy and sensory loss are found in addition to pain and tenderness. If it occurs in one limb it is usually because of some gross lesion such as pressure on, or irritation of, a particular nerve trunk.

Thoracic or cervical outlet compression syndrome

CAUSES	VASCULAR FEATURES	NEUROLOGICAL FEATURES		
		Afferent	Efferent	Sympathetic
Weak shoulder girdle muscles. Cervical rib or band	Neck – subclavian artery bruit Hand – Raynaud's phenomenon, positive 'Adson's' test	Pain radiates down inner side of arm Diminished sensation in C8 dermatome	Intrinsic muscle weakness in T1 myotome	'Raynaud's' phenomenon, 'Horner's' syndrome

766. Cervical rib syndrome. This is characterised by an aching, dull, boring pain which radiates from the root of the neck down the inner side of the arm and forearm to the tips of the fingers. It may be associated with an ache in the shoulder and scapula, but it does not radiate into the head. Women suffer more than men. It may be worse at night after a day's work or if the patient sleeps on the affected side.

There may be associated motor disturbances (intrinsic muscle palsy), sensory disturbances (anaesthesia along the ulnar border of the forearm and little finger) and vasomotor disturbances (Raynaud's phenomenon). There may be diminution of the radial pulse on the affected side.

766

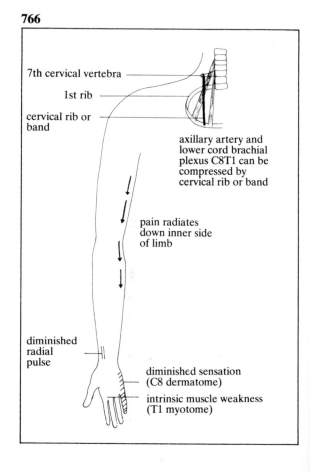

7th cervical vertebra
1st rib
cervical rib or band

axillary artery and lower cord brachial plexus C8T1 can be compressed by cervical rib or band

pain radiates down inner side of limb

diminished radial pulse

diminished sensation (C8 dermatome)

intrinsic muscle weakness (T1 myotome)

767

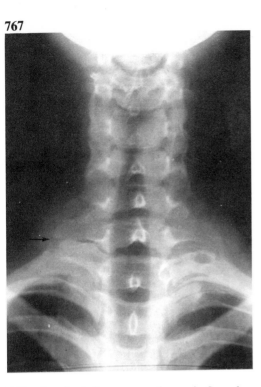

768

769

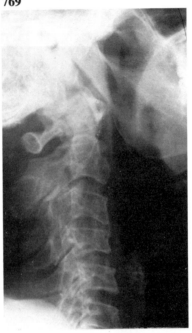

767. Cervical rib on x-ray (arrow). See also **676**.

768. Cervical spondylosis. This 55-year-old man had pain in the C5-6-7 dermatomes of his right upper arm. He had weakness and wasting of the brachioradialis muscle (arrow).

769. Cervical spondylitis with narrowed intervertebral foramina C5-6-7.

770. Intervertebral disc protrusion. Chronic pain in the C5-6 dermatomes was relieved by disc excision and cervical fusion (see scar). The thenar muscle wasting persisted. He was previously misdiagnosed as having carpal tunnel syndrome. Surgical decompression had not relieved his symptoms.

771. Myelogram. Disc protrusion at C5-6 level.

770

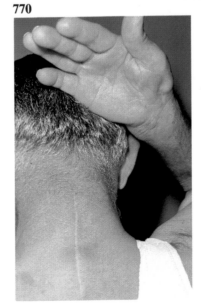

771

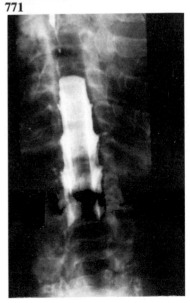

Referred pain from the thorax

Patients with ischaemic heart disease may complain of paroxysms of acute pain radiating in the area of the 1st and 2nd dorsal nerve roots down the ulnar border of the arm and sometimes into the little finger. Cutaneous hyperaesthesia may be present in the same areas. In all cases of paroxysmal pain referred to the left arm, a careful examination of the thoracic viscera is indicated.

Pancoast's syndrome. This is the result of carcinoma beginning in the apex of the lung and pressing on neighbouring structures. The syndrome comprises distension of the veins of the neck from pressure on the superior vena cava, swelling of the face from the same cause, Horner's syndrome from pressure on the sympathetic chain, shooting pains down the arm and later a lower brachial plexus lesion.

772–774. Pancoast tumour with Horner's syndrome and wasting of the small muscles of the hand. Carcinoma of the apex of the right lung.

772

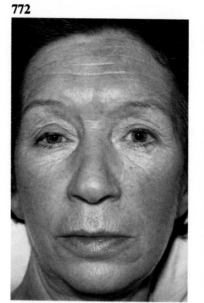

773

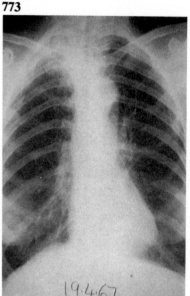

19.4.67.

774

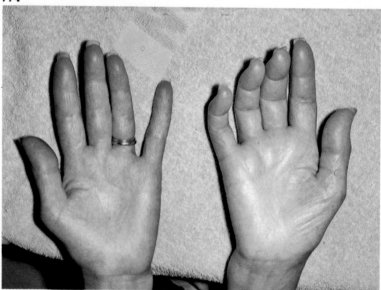

Henry Khunrath Pancoast (1875–1939). American radiologist. Pancoast syndrome. A syndrome observed in carcinoma of the apex of the lung (Pancoast's tumour), characterised by local pain irradiating towards the shoulder, arm, and hand; sensory and motor disorders and wasting of the muscles of the hand; rib lesions; and Bernard Horner syndrome.

Miscellaneous

Hysteria, neuraesthenia

These patients may present with acute pain in the hand or arm. The characteristic features are: absence of a definite pain pattern, the presence of physical signs without an underlying anatomical basis. There may also be an association with other hysterical or neuraesthenical conditions such as functional aphonia, globus hystericus, stocking and glove anaesthesia, hemi-anaesthesia, hysterical seizures, variable paralyses from the contraction of antagonistic muscle groups. Hysteria is commoner in women and neuraesthenia commoner in men.

775. Hysterical pain and disuse after minor injury to the pulp of the right thumb. This woman had anaesthesia involving her entire hand, i.e. glove type anaesthesia.

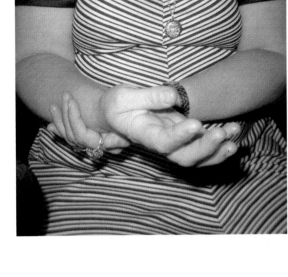
775

Hysteria. *Hyster* in Greek means a uterus. Hysteria literally means a uterine condition, so called by the ancient Greeks who believed that the nervous symptoms were due to the uterus and were thus observed only in women.
    In 1682 Thomas Sydenham (the 'English Hippocrates', 1624–1689) contributed the classic account of hysteria.

Occupational neuroses

The hand and upper limb is the commonest site of occupational neuroses, e.g. writer's cramp or typist's cramp, according to whether the occupation is writing, typing, needlework, hair cutting etc. There is usually some form of muscle spasm. The pain and spasm are evolved by the employment of the limb in a particular occupation, but other manipulations using the same muscles may be carried out with impunity. The acute pain which is associated with the spasm may be followed by a dull aching pain for some hours afterwards.

9. Occupational conditions

Skin

Occupational stigmata
callosities, hyperkeratoses and telangiectases
 776–783
stains 784–786
scars 787–792
colour changes 793–795
knuckle pads 796
barber's sinus 797

Occupational dermatitis
contact dermatitis 798
arsenical keratoses 799
cement dermatitis 800

Occupational infections
erysipeloid 801
anthrax 802
orf 803
tuberculosis 804, 805

Miscellaneous
radiation necrosis 806

Nails
paronychia 807
onychomosis 808
acute ungual eczema 809
long fingernails 810, 811

Fascia and bursa
Dupuytren's contracture 812–813
occupational bursitis

Tendon
traumatic and occupational disorders of
 tendons 814
myotendinitis 815
flexor tenosynovitis 816
trigger finger 817
extensor tenosynovitis 818, 819
de Quervain's syndrome 820, 821

Nerves
carpal tunnel syndrome 822–824
ulnar neuropathy 825, 826
traumatic neurofibroma 827–829

Blood vessels
hand hammer syndrome 830
digital hammer syndrome 831, 832

Skeleton
bone cysts 833
game-keeper's thumb 834
osteo-arthritis 835

Miscellaneous
writer's cramp

Some occupations are associated with a harmless stigma such as a timber stain, some show a direct cause/effect relationship such as callus formation over pressure points, and others aggravate a pre-existing tendency, e.g. to develop osteo-arthritis or Dupuytren's contracture.

Occasionally injuries and infections occur as an accident during the course of man's work.

Occupational disorders and effects in the hand can be classified as follows:

(1) Aetiologically	*(2) Anatomically*
e.g. vibration	skin
exercise	nail
pressure	fascia and bursa
friction	tendon
injury	nerve
infection	blood vessels
radiation	skeleton
psychological	miscellaneous

Skin

Occupational stigmata

Nearly every occupation or handicraft manifests itself on the hands as distinguishing callosities, hyperkeratoses, telangiectases, stains, scars, colour changes, knuckle pads, sinuses and deformities.

Callosities, hyperkeratoses and telangiectases

776. Callosities and dirt stains on the palmar skin and thumb of a labourer. Compare the relative lack of such calluses and stains in an office worker.

777. Disuse (trophic) skin on the left thumb pulp after nerve injury. The right thumb pulp shows normal work staining, i.e. ingraining of dirt.

776

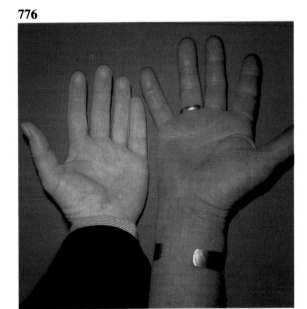

777

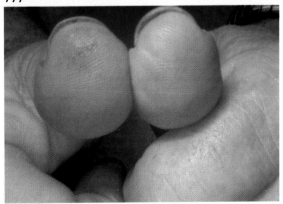

778

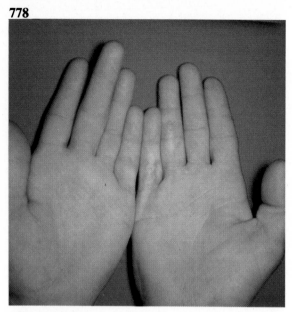

779

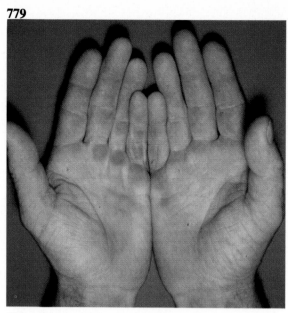

778. Callosities over the bony prominences of the right ring and little fingers from prolonged friction and pressure of a butcher's knife.

Racquet players develop calluses over the gripping surface on the ulnar side of the hand.

779. Grease stains and callus on the gripping surfaces of both hands in a fitter working in an engine room.

780

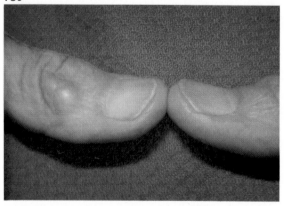

781

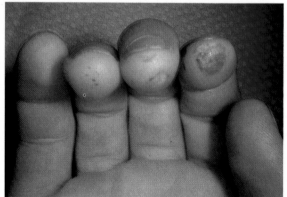

780. Callus in a clarinetist. A clarinetist develops a callus over the dorsum of the right thumb from pressure of the hook used to support the instrument.

781. Fingertip callosities in a double bass player.

782

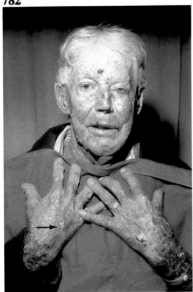

783

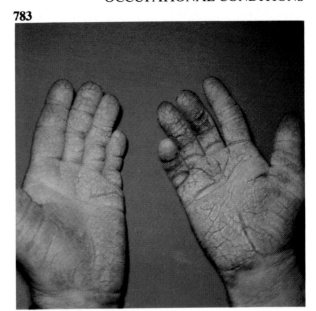

782. Solar keratoses and telangiectases (arrow) of exposed skin in a farmer. Sailors, fishermen and other outdoor workers also develop similar stigmata.

783. Thick fissured palmar skin in a fisherman.

784

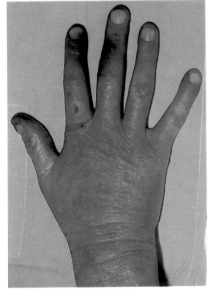

785

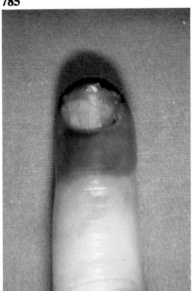

Stains

784 & 785. Nicotine stains in a heavy smoker. Sixty cigarettes a day produces obvious stain; even 5–10 cigarettes a day can be smelt on the fingers. Staining of the palm may occur in cigarette smokers, particularly if the cigarette is held with the ignited end between the fingers pointing into the palm where the smoke can drift out of the hollow and stain it.

786. Timber sap stains in a saw miller. Note absence of stain on the right middle finger which was previously injured and is no longer used.

Stains may also be seen in the skin and nails of printers and photographers.

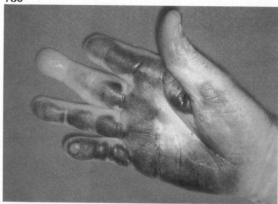

Scars from occupational injury, infection, foreign body and burns

787. Blue metal sores. Recurrent fingertip and nail sores in a road builder whose job involved the handling of blue metal.

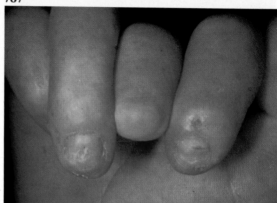

788 & 789. Lime sores. Recurrent sores from lime irritation in a concrete worker.

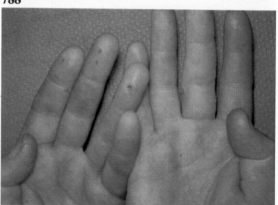

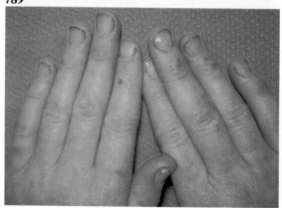

790. A butcher's hand. Arrows show scars from a recent and from previous slicing injuries.

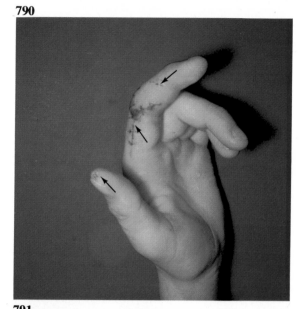

791. Numerous superficial cuts in a glazier.

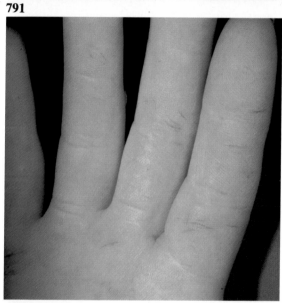

792. Indentation of the left thumbnail in a cabinet-maker who used a knife to trim pieces of wood held between his thumb and index finger.

A watchmaker's right thumbnail is often short and hard and hypertrophied from the constant opening of watches.

Burns from sodium hydroxide, silver nitrate or liquid chlorine may be seen in the hands of chemical workers.

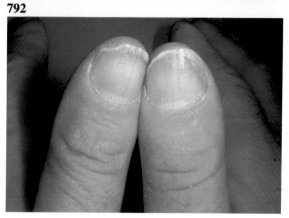

Colour changes in the skin. These sometimes characterise an occupation.

793 & 794. Ingrained coal dust in the right thumb of a collier. X-rays show the coal fragments.

793

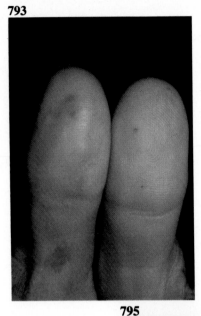

794

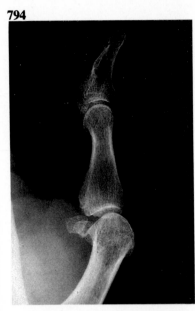

795. A printer at work with his stencils. Note the ink stains on his index finger and thumb.

795

796. Knuckle pads in a shearer.

797. A barber's sinus. Hair embeds in the interdigital cleft.

796

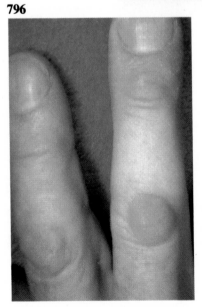

797

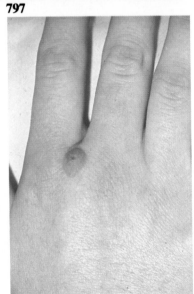

Occupational dermatitis

This can be caused by:

(1) Physical factors – pressure, abrasion, moisture, desiccation, heat, cold, light, x-rays and other rays.

(2) Plant products – leaves, stems, sap, roots, bulbs, flowers, fruits, vegetables, wood dusts, resins, lacquers.

(3) Living agents – bacteria, viruses, fungi, helminth parasites, insects and mites.

(4) Chemical substances – inorganic acids and salts, hydrocarbons, oils, tar, pitch, anthracene, and dyes.

798. Contact dermatitis, i.e. 'dermatitis venenata', 'occupational or industrial dermatitis'. This is characterised by redness, oedema, vesicles and itching, and is caused by chemical or vegetable substances contacting the skin.

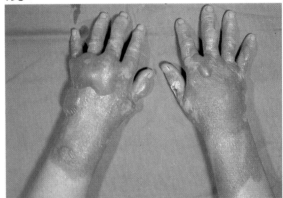

799. Arsenical keratoses in a farmer from handling liquid arsenic. Arsenic can cause pigmentation, palmar and plantar hyperkeratoses and eventually intra-epidermal carcinoma. These hands show palmar hyperkeratosis. There is prominent nicotine staining of his left hand.

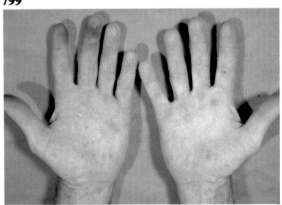

800. Cement dermatitis.

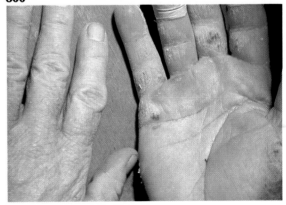

Occupational infections

Persons whose hands contact infectious animals are liable to potentially serious infections. Always ask for, in any infection of the hand, a history of occupational contact.

Occupational disease due to infections

INFECTION	ORGANISM	ANIMAL VECTOR	CLINICAL LESION
Erysipeloid, 'fish finger'	Erysipelothrix	Fish, cattle	Acute inflammatory induration of the skin and soft tissues. Less pain and systemic reaction than in an acute infection
Anthrax, 'malignant pustule'	Gram positive spore bearing aerobic bacillus	Sheep, cattle, horses	Papule → pustule → ulcer
Brucellosis	Gram negative aerobic rods	Cattle, hogs, goats	Lymphadenitis and lymphadenopathy
Milker's nodule	Vaccinia virus	Cows	Pruritic papules → blue red nodules → ulcer
Glanders	Pseudomonas mallei Gram negative aerobic rod	Horses	Ulcers, lymphadenitis
Tularaemia	Gram negative aerobic rod	Rodents, insects	Papule → ulcer → lymphadenitis
Tuberculosis	Mycobacterium Tuberculum	Cattle	Papule → ulcer Tenosynovitis
Orf, 'pustular' dermatitis	Virus	Sheep, goats	Papule → pustule

801. Erysipeloid (fish handler's disease). Occurs almost exclusively in those handling fish and meat. After a scratch or cut, the area becomes inflamed or red or purple in colour. The infection lies in the subcutaneous tissue. The constitutional reaction is relatively slight.

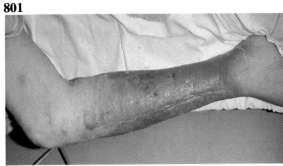
801

802. Anthrax. It is also known as malignant pustule but is rarely fatal. A gangrenous carbuncular lesion in the left index finger of a cattle worker.

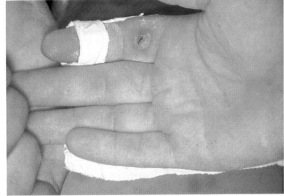
802

803. Orf. Three to seven days after inoculation from a contagious sheep or goat a firm painless dark papule may appear on the finger or hand. A pustule may then develop. The condition is self-limiting, clearing in 4–8 weeks.

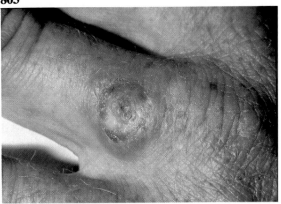
803

804. Tuberculosis. 'Lupus vulgarus'. Direct inoculation of the tubercle bacillus during slaughter of an infected calf resulted in this subcutaneous nodule. Such nodules can grow and ulcerate.

805. Tuberculosis. This abbatoir worker had a carcass bone puncture in his hypothenar eminence. He developed tuberculous tenosynovitis and a chronic discharging sinus.

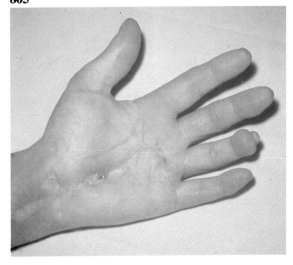
805

804

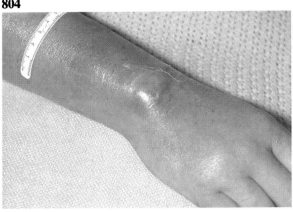

Miscellaneous

806. Radiation necrosis. Over exposure to x-ray resulted in radiation necrosis and ischaemia necessitating amputation of the right index and ring fingers.

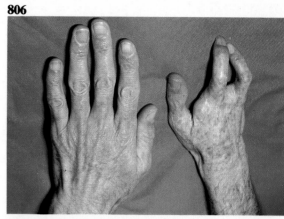
806

Nails

Irritations, inflammation and infection. Chronic paronychia occurs in those continually immersing their fingertips in water, e.g. barmaids.

807. Chronic paronychia in the left thumb. There is thickening of the nail folds and secondary nail deformity.

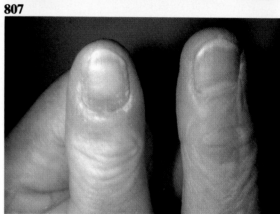
807

808. Onychomycosis. This occurred from brewer's mould found in fermentation vats.

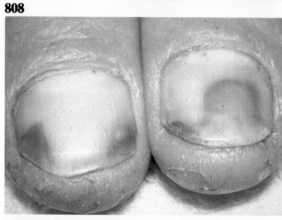
808

809. Acute ungual eczema can be found in those handling lime, nickel, salts, and formaldehyde.

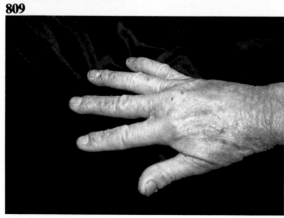
809

Stains and scars. The nail, like the skin, can show occupational stigmata such as stains (red nails in vineyard workers, dark brown nails in photographers), and wear (in watchmakers and those working grinding machines).

810. Guitarist's fingernails are kept long to help them play their instruments.

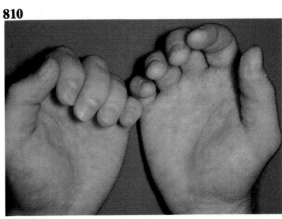

810

811. Digit auricularis. Some races keep their left little fingernail long to assist in cleaning the ear.

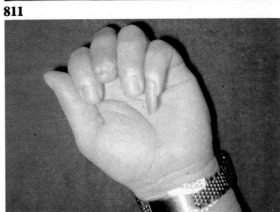

811

Fascia and bursa

Palmar fasciitis like Dupuytren's palmar fibrosis has been observed in those handling vibrating tools and, for example, upholsterers and oarsmen whose hands are subject to repeated blows and friction.

812 & 813. Dupuytren's fibrosis and contracture in the right hand of a waterside worker who for 30 years grappled bales of wool with a hook.

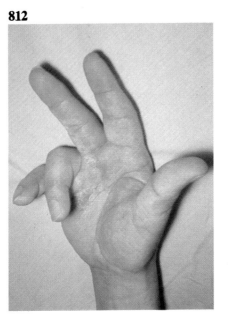

812

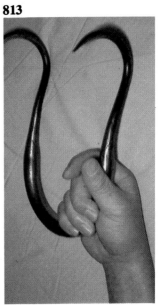

813

Occupational bursitis (miner's beat hand). Pressure, friction or repeated blows can inflame a pre-existing bursa (olecranon bursitis) or generate an adventitious bursa, e.g. in the subcutaneous tissues of the hypothenar eminence.

The subcutaneous tissues of the hands become thickened in coalminers and stokers from prolonged pressure of a pick or shovel. Infection of these tissues, often from staphylococcal infection of a hair follicle, gives rise to subcutaneous cellulitis which is called a 'beat hand'. This infection may extend deeply to involve the tendon sheaths.

Tendon

814. Traumatic and occupational disorders of tendons. These are probably the commonest occupational diseases affecting the hand. There are five sites. All these conditions represent non-infective inflammation of the tendon apparatus, caused by unaccustomed and arduous use of the hand. Chicken workers who pluck feathers from hens, and others whose work involves a pincer action between the thumb and fingers, accompanied by quick pronation and supination of the forearm are most likely to develop the condition. It is usually unilateral. Clinically the patient presents with localised pain, swelling, tenderness, fine crepitus and restriction of function. Coarse crepitus probably indicates rheumatoid or tuberculous tenosynovitis.

814

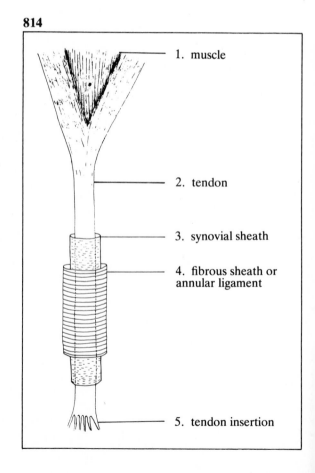

1. muscle

2. tendon

3. synovial sheath

4. fibrous sheath or annular ligament

5. tendon insertion

Tendinitis, tenoperiostitis and musculotendinitis. Excessive strain may cause rupture of fibres with the substance of a tendon at three sites – at the muscle tendon origin, within the substance of the tendon, or at the tendon insertion to bone.

The resultant scar can remain persistently painful and will be aggravated by muscle action.

815. Myotendinitis (arrow). Pain, redness, swelling, tenderness and crepitus of the wrist extensor tendons after strenuous exercise.

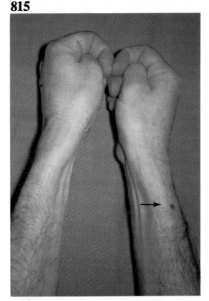

Occupational tenosynovitis. Excessive or forced repetitive movements of a tendon within its sheath can cause roughening of the external aspect of the tendon and the internal aspect of the tendon sheath. Gliding of the surfaces one against the other causes pain and fine crepitus.

816. Flexor tenosynovitis with pain, swelling and crepitus in the right wrist (arrow) of a 45-year-old female process worker.

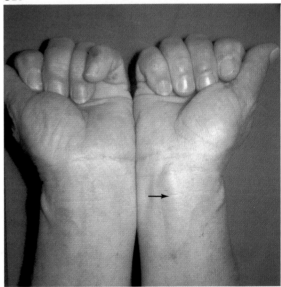

817. Trigger finger, 'locked finger', 'clicking finger'.

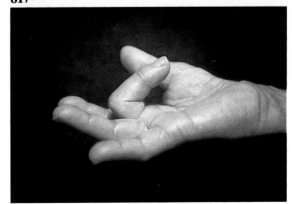

818

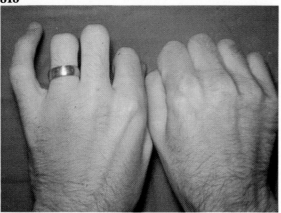

818. Extensor tenosynovitis of the left ring finger in a guitarist who played non-stop for two days. He had painful swelling and crepitus.

819

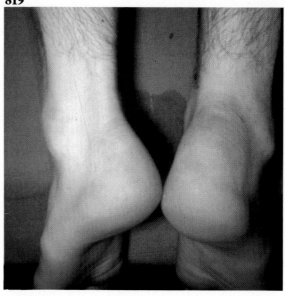

819. Associated tendonitis of the right Achilles' tendon from repetitive pumping of the drum pedals.

820

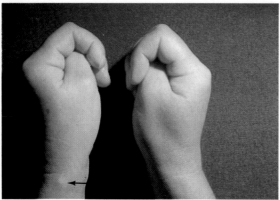

820. Tenovaginitis ('de Quervain's syndrome'). This is a primary thickening of the fibrous sheath which may follow repeated strain or blunt trauma. The most frequent example is de Quervain's syndrome which occurs over the radial side of the wrist where the long abductor and short extensor tendons of the thumb pass through the most radial sheath. There may be secondary irritation of the overlying superficial branch of the radial nerve.

821

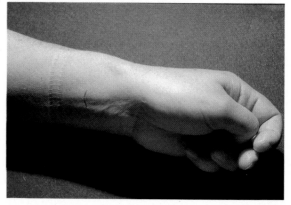

821. Finkelstein's test. Full active flexion of the three joints of the thumb with attempted ulnar deviation of the wrist produces pain over the radial side of the wrist.

Nerves

822. Carpal tunnel syndrome. Compression median neuropathy at the wrist. Any increase in the content of the carpal tunnel may compress the median nerve. The commonest cause is proliferative flexor tenosynovitis which can be associated with occupations involving constant repetitive forced actions of the wrist and fingers.

It can also occur in occupations involving repetitive use of the hand with the wrist extended, so causing the median nerve to grind against the carpus, e.g. in using clippers or in scrubbing the floor.

In both instances it is thought that the occupation is an aggravation rather than an actual cause of the condition.

823 & 824. Bilateral carpal tunnel syndrome in a 60-year-old shearer. For 40 years he used both hands in forced repetitive manner. He tolerated the pain of carpal tunnel syndrome but noticed increasing weakness of pinch, especially on the left side. Wasting, particularly of the left thenar eminence is seen in the lateral view.

822

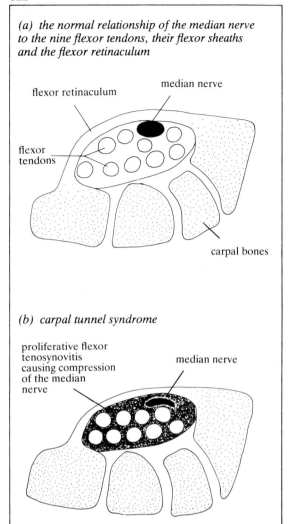

(a) the normal relationship of the median nerve to the nine flexor tendons, their flexor sheaths and the flexor retinaculum

flexor retinaculum

median nerve

flexor tendons

carpal bones

(b) carpal tunnel syndrome

proliferative flexor tenosynovitis causing compression of the median nerve

median nerve

823

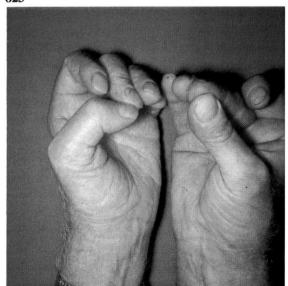

824

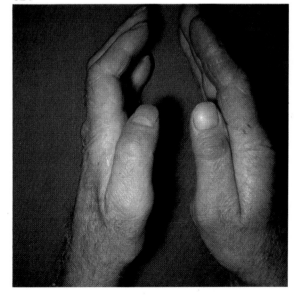

Ulnar neuropathy. This can occur at the elbow or at the wrist. Occupations requiring repetitive elbow movements can in some individuals cause friction of the ulnar nerve in the groove behind the medial epicondyle and give rise to sensory and motor symptoms in the hand. Always compare both elbows, exclude a skeletal abnormality, and test for diabetes, a common cause of mononeuritis.

825. Ulnar neuritis with claw deformity and intrinsic muscle wasting in a 36-year-old labourer. He spent his working life using a crowbar and shovel. Repetitive elbow movements caused frictional neuritis of the ulnar nerve at the elbow.

826. Wasting of intrinsic muscles seen in the dorsal view. Prolonged pressure or hammering with the hypothenar eminence can be associated with ulnar neuritis at the wrist. See Chapter 3.

At both elbow and wrist the ulnar nerve can be compressed by a ganglion. See Chapter **718**.

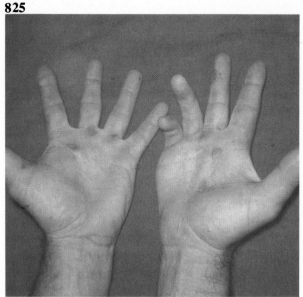

825

826

ganglion compressing
ulnar nerve

ulnar blood
vessels

median
nerve

ulnar
nerve

carpal
tunnel

Neurofibroma. Repeated pressure on a nerve can cause a neurofibroma.

827. 'Bowler's thumb'. A neurofibroma can develop on the ulnar side digital nerve from repeated pressure on the nerve during bowling. Bowlers can also develop thickening of the interphalangeal joint of the thumb.

827

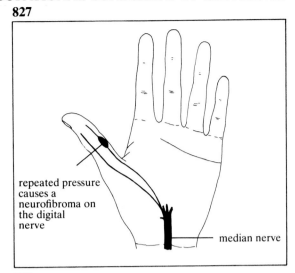

repeated pressure causes a neurofibroma on the digital nerve

median nerve

828. Neurofibroma of the dorsal branch of the digital nerve in a 30-year-old baker. Repetitive pressure and friction applied to the left middle finger during the handling of a baker's tray produced this swelling.

828

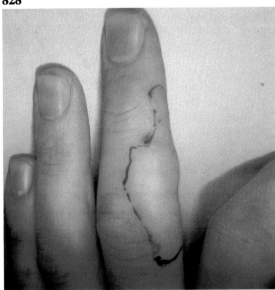

829. Operative finding.

829

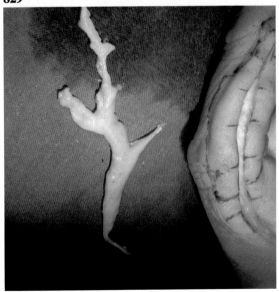

Blood vessels

Vascular occupational disorders include dead hand, dead fingers, white fingers, pneumatic hammer disease, pseudo Raynaud's disease, traumatic vasospastic disease, vibration induced white fingers.

Vibrating tools are incriminated as a cause of Raynaud's phenomenon, repeated blows to vessels in the hand being associated with the hand hammer or digital hammer syndrome.

The Raynaud's phenomenon is defined by Hunt (1936) as intermittent pallor or cyanosis of the extremities precipitated by exposure to cold, without clinical evidence of blockage of the large peripheral vessels. The occasional presence of ischaemic lesions is limited to the skin.

Maurice Raynaud (1834–1881). French physician. Described local asphyxia and symmetrical gangrene, since known as Raynaud's disease, in 1862.

830. The hand hammer syndrome. This 60-year-old carpenter used his hypothenar eminence as a hammer. His little finger became cold, blue and numb. The physical findings of diminished blood supply included diminished capillary filling, absent digital pulses and tenderness over the ulnar artery at the wrist.

830

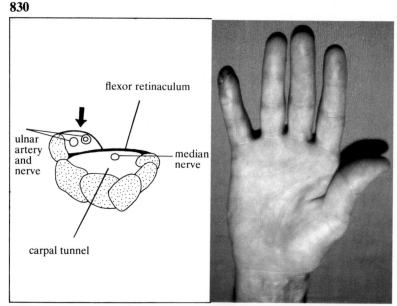

831. Digital hammer syndrome. Chronic ischaemic ulcer of the right middle finger from repeated striking of the base of the finger against a heavy spanner.

832. Poor perfusion of the digital blood vessels to the right middle finger.

831

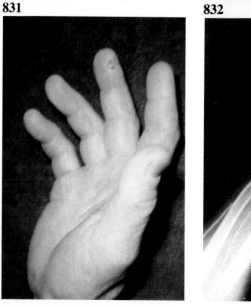

832

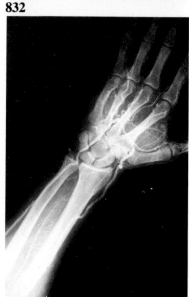

Skeleton

833. Small areas of decalcification or cysts seen in the x-rays of the bones of the carpus. The prolonged use of vibrating tools may be associated with this. These signs are also seen in other individuals engaged in heavy manual work. They are symptomless and do not predispose to fracture or any other complication.

Vibration tools are also associated with osteoarthritis of the joints of the hand and wrist. The prolonged use of cutting instruments by hair dressers, shearers and tailors predisposes to osteoarthritis those joints handling the instruments.

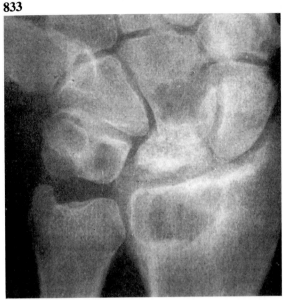

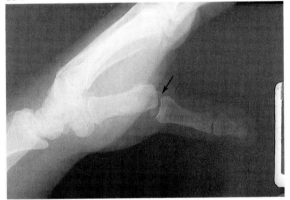

834. Gamekeeper's thumb. Keepers who use their thumb to wrench and break the necks of small animals are prone to rupture the ulnar collateral ligament (arrow) of the metacarpo-phalangeal joint.

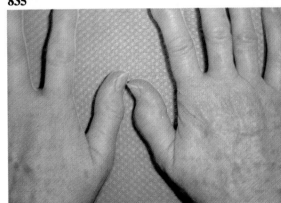

835. Osteo-arthritis of the interphalangeal joint of the thumb in a tailor.

Miscellaneous (writer's cramp)

This is an occupational neurosis, characterised by painful spasms of the hand and forearm, as one attempts to write. The mental concept of writing interferes with the act of writing although other activities requiring fine motor co-ordination, such as the tying of shoe laces, can be done with ease. There are no abnormal physical signs.

The patient may feel a cramp or pain as soon as he grasps a pen. The delayed onset of pain may be brought about by ischaemic contraction of the forearm muscles.

Differential diagnosis. (1) Painful writer's cramp should be distinguished from the carpodigital spasm of hypoventilation tetany. (2) Writer's cramp can usually be distinguished from 'paralysis agitans' – in this latter condition quick rotation of the forearm to and fro cannot be performed on the affected side. (3) Nerve entrapment – carpal tunnel syndrome.

10. Systemic disease involving the hand

physical signs **836**

Disorders of connective tissue
osteo-arthritis **837–840**
rheumatoid arthritis **841–852**
psoriasis **853–856**
Dupuytren's contracture **857–864**
scleroderma **865–868**
systemic lupus erythematosis **869, 871**
polymyositis and dermatomyositis **872**
sarcoidosis **873**
Reiter's syndrome **874**
Marfan's syndrome **875**
Ehlers-Danlos syndrome **876–878**
epidermolysis bullosa **879, 880**

Diseases of the cardiovascular and respiratory system
tetralogy of Fallot **881**
bacterial endocarditis **882–884**
myocardial infarction
clubbing of the fingers **885, 886**
hypertrophic pulmonary osteo-arthropathy **887–890**

Disorders of metabolism
gout **891–900**
hepatic cirrhosis **901–907**
diabetes mellitus **908–910**
rickets **911**
porphyria **912, 913**

Disorders of the nervous system 914
syringomyelia **915**
Parkinson's disease **916**
spastic conditions **917, 918**
neuralgic amyotrophy **919**
amyotrophic lateral sclerosis **920**
myotonia **921–923**
hysteria **924**
peripheral neuritis **925–927**
mongolism **928**
hyperhidrosis **928**

Disorders of the endocrine system
acromegaly **929–932**
hypothyroidism **933, 934**
hyperthyroidism **935, 936**
hypoparathyroidism **937, 938**
hyperparathyroidism **939**
Addison's disease **940–942**

Haemopoietic diseases
anaemia **943, 944**
polycythaemia **945–947**
purpura **948–950**

Miscellaneous disorders
palmar erythema **951**
oedema of the hand **952**
senile changes in the hand **953**
Peutz-Jeghers' syndrome **954, 955**
pigment changes **956, 957**

No organ, anatomic structure or laboratory procedure can reveal so much practical information about the patient so readily as can the hand; nor is there any other structure so convenient and accessible for examination.

Systemic diseases can manifest in one or more component tissues of the hand. Signs in the hand sometimes appear years before those of the systemic disease.

In some conditions, for example dermatomyositis, the hand signs (skin rash and arthralgia) are an indication of developing internal malignancy; e.g. cancer of the alimentary tract.

The physical findings in the hand may give many clues towards the diagnosis of a systemic disease.

Examine the hand (the palm, the dorsum and the nails separately) by: Inspection, Palpation, and Auscultation.

836. Some physical signs to be found on examination of the palm, dorsum and nail.

836

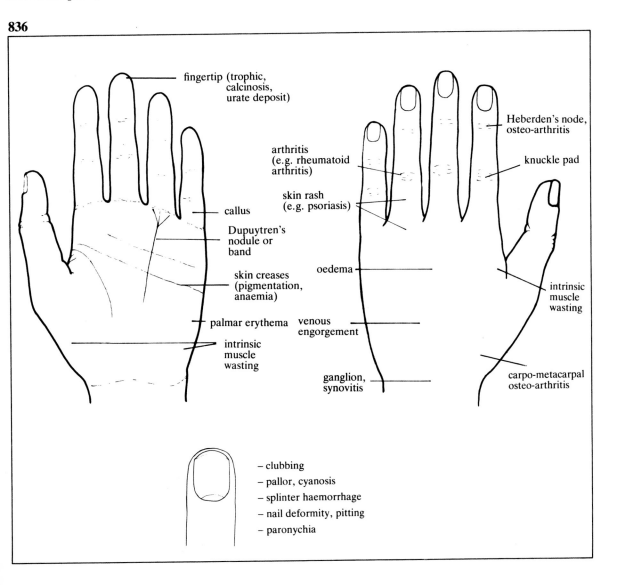

Inspection

Size and shape, e.g. acromegaly, oedema etc.

Deformity, e.g. rheumatoid arthritis, osteo-arthritis, gout etc.

Position, e.g. wrist drop.

Colour of skin, e.g. anaemia, jaundice, Addison's disease.

Texture of skin, e.g. scleroderma.

Pulsations, e.g. arteriovenous fistula.

Tremor, e.g. anxiety, writer's cramp, Parkinson's disease, thyrotoxicosis, disseminated sclerosis.

Haemorrhages, e.g. purpura.

Lumps, e.g. ganglion, xanthomas etc.

Scars, e.g. injury, inflammation, burns.

Muscle wasting, e.g. nerve injuries.

Movements, e.g. hemiplegia.

Nails, e.g. clubbing, cyanosis, splinter haemorrhages, pitting.

Palpation

Confirm the findings of inspection.

Temperature, e.g. increased general temperature of the hand in fever, thyrotoxicosis etc, and increased local temperature from arteriovenous fistula. Decreased general temperature in hypothyroidism, shock etc. and decreased local temperature in peripheral vascular disease.

Power is diminished after nerve palsy.

Sensation is diminished after nerve injuries.

Pulsations are seen in arteriovenous fistulae.

Sweating is increased in anxiety and thyrotoxicosis.

Auscultation

For a bruit, as in arteriovenous fistulae.

Disorders of connective tissue

Osteo-arthritis

This is perhaps the most common systemic disturbance that is seen in the hands. 90% of people over 40 years have x-ray evidence of osteo-arthritis and about 5% of persons have symptoms from it. Women are afflicted more often than men.

There are three different types.

(a) Wear and tear causes osteo-arthritis of the distal interphalangeal joints of the fingers and the carpo-metacarpal joints of the thumb in the middle and older age groups.

(b) Occupational stresses may affect particular joints, e.g. the joints of the thumb and index finger in a hairdresser using scissors.

(c) Injury, especially an intra-articular fracture, may predispose a joint to osteo-arthritis.

Pathogenesis. Wear and tear erodes the articular cartilage, thus narrowing the joint space. Adjacent bones become dense and eburnated from contact with their fellow, causing cystic erosions at the bone ends. Bone spurs, osteophytes, and exostoses form around the bony margins of the joint. These are called 'Heberden's nodes'. Secondary synovial inflammation occurs and may be responsible for ganglia or mucous cysts arising from the distal interphalangeal joint. See Chapter 5.

The clinical features include: pain, swelling, stiffness and deformity.

William Heberden (1710–1801). English physician. A Greek and Hebrew scholar. He wrote his case notes in Latin – these, his 'commentaries', were published after his death in 1802 and translated into English by his son. *Heberden's nodes.* Hard nodules adjacent to the distal interphalangeal joints of the fingers, occurring in osteo-arthritis.

837. **Osteo-arthritis** of the distal interphalangeal joints with a characteristic Heberden's node on the dorsal aspect of the joint. Flexion and deviation deformity, especially of the right index and middle finger.

The x-ray features of osteo-arthritis are diminished joint space, subchondral bone sclerosis and osteophyte formation.

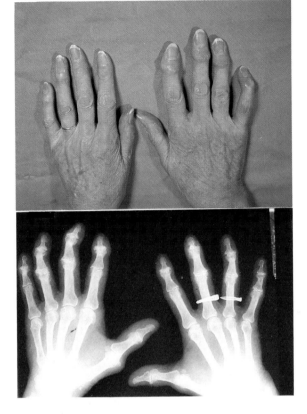

838. **X-rays of osteo-arthritis** of the interphalangeal joint of the right thumb in a hairdresser.

839. **Osteo-arthritis** and secondary subluxation of the carpo-metacarpal joint of the right thumb. Secondary hyperextension of the metacarpophalangeal joint (see contractures).

840. **X-ray.**

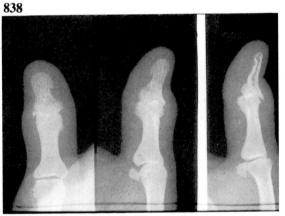

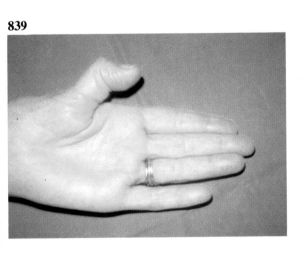

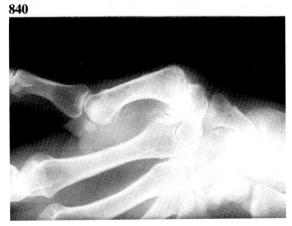

Rheumatoid arthritis

Rheumatoid arthritis is a synovio-arthritis affecting the proximal finger joints, e.g. the metacarpophalangeal and proximal interphalangeal joints, although it can affect any part of the hands where there is synovium. (Osteo-arthritis affects mainly the distal interphalangeal joints.)

In rheumatoid arthritis, inflammatory synovial pannus spreads over and erodes the articular cartilage, reducing the joint space and eroding the underlying bone. It distends the joint capsule and ligaments. The joint later dislocates. Direct involvement of muscles and tendons causes secondary contractures. Inflammatory swelling and oedema affects the extra-articular soft-tissues, giving the joints a spindle-shaped appearance. Tendons whose synovium is involved may show nodules, triggering, dislocation and rupture.

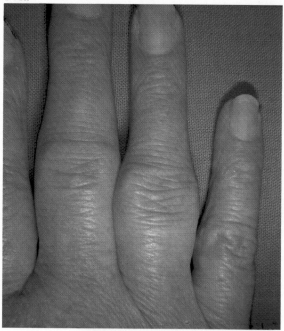

841a. Early rheumatoid arthritis. Spindle-shaped swelling of the proximal interphalangeal joint.

841b. X-ray of early rheumatoid arthritis. There is soft-tissue swelling, mainly affecting the middle finger, giving it a spindle-shaped appearance. There is narrowing of all proximal interphalangeal joints. The proximal interphalangeal joint of the middle finger shows cystic erosion at the base of the middle phalanx, with marginal erosion of the distal end of the proximal phalanx.

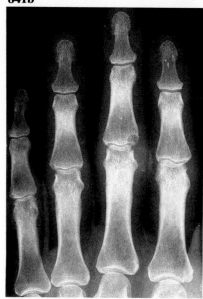

842a & 842b. Rheumatoid arthritis with endarteritis, causing an ulcer at the distal interphalangeal joint and haemorrhage at the nailfold (arrow).

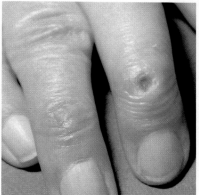

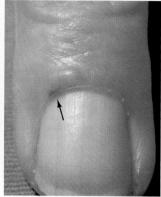

843

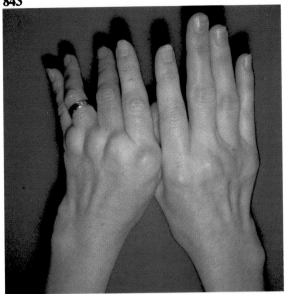

844

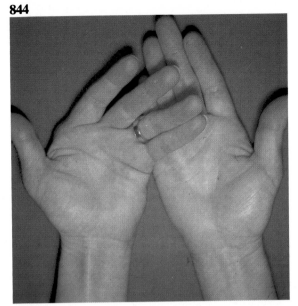

845

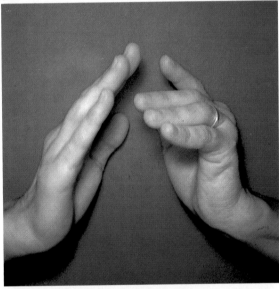

843–845. Rheumatoid arthritis of the metacarpo-phalangeal joints, more severe on the left hand with proliferative synovitis, ulnar deviation, and metacarpo-phalangeal dislocation.

846

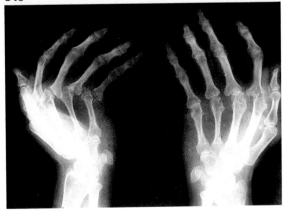

846. X-ray features showing a diminished joint space, absorption of bone just beneath the cortex adjacent to the joint, ulnar deviation of the fingers, and metacarpo-phalangeal dislocation (on the left hand).

847

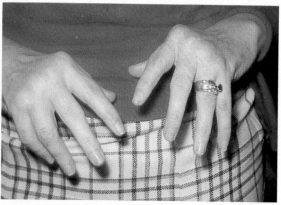

848

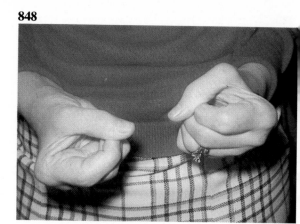

847 & 848. Rheumatoid arthritis with metacarpophalangeal joint dislocation; fibrosis of the intrinsic muscles prevents full flexion of the fingers.

849

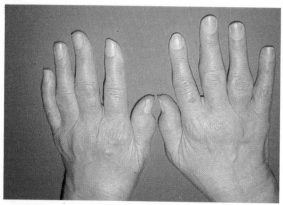

849. Rheumatoid synovitis involving the extensor tendons at the right wrist. The metacarpophalangeal joints are also involved.

850

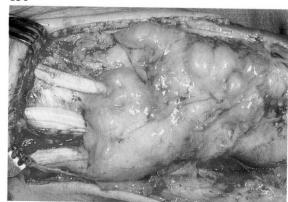

850. Operative findings. Extensor tenosynovitis.

851

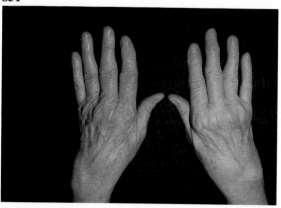

851. Rheumatoid arthritis with intrinsic contracture (swan neck) of the proximal interphalangeal joint of the right ring finger.

852

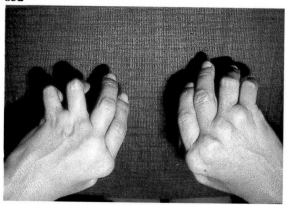

852. Rheumatoid arthritis with closed extensor tendon ruptures, from friction over the roughened proximal interphalangeal joints of the left ring and little fingers.

Psoriasis

Psoriasis is a multi-system disease manifesting in the hand as changes in the skin, joint and nail.

The skin rash is a dry, whitish scale, irregularly covering a dull, red, sharply outlined base. It is situated usually on the extensor surface of the body particularly of the skin overlying involved joints.

The first joint signs are peri-articular swelling of the interphalangeal joints.

The nails are thick, irregular and show pitting.

853. **Psoriatic arthritis** of the distal interphalangeal joints and the typical skin rash of the abdomen.

854. **Psoriatic arthritis** of the proximal interphalangeal joint of the right ring finger with skin rash over the interphalangeal joint.

855. **Psoriatic arthritis.** An advanced case in a Chinese male. Differential diagnosis: rheumatoid arthritis.

856. **X-ray of a similar case.** Gross destruction with subluxation and ankylosis of the interphalangeal joints.

853

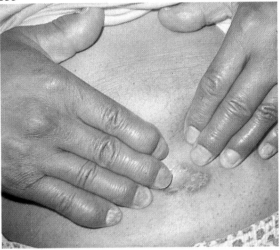

854

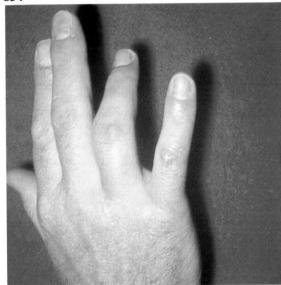

856

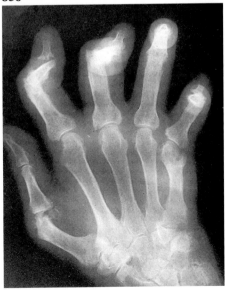

855

Dupuytren's contracture

Dupuytren's contracture of the palmar and digital fascia was first described by Guillaume Dupuytren in 1831. The cause is unknown and it is primarily a disease of Caucasian males aged 50–70, being rare in Orientals and Negroes. There is a positive family history in 25% of cases, and it occurs as readily in sedentary as in manual workers. It has a higher incidence in alcoholics, diabetics and epileptics.

Primarily it involves the longitudinal bands of the palmar fascia, the vertical extensions of that fascia to the skin and the oblique and horizontal fascia of the webs.

It is associated with knuckle pads, plantar fasciitis and penile fasciitis (Peyronie's disease).

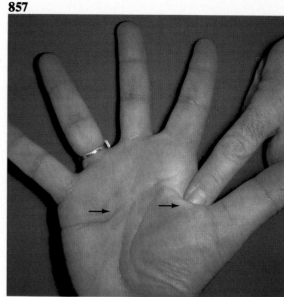

857. Early Dupuytren's contracture. Thickening of the palmar fascial band of the ring finger with early skin tethering, puckering and pit (arrow) formation. No joint contracture. Involvement of the fascial band in the web space between thumb and index finger.

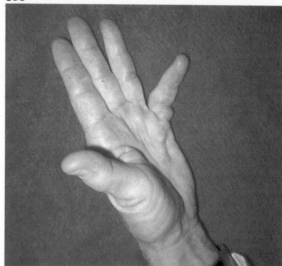

858. Progressive painful Dupuytren's fasciitis with enlarging Dupuytren's nodules, contracture of the little finger, and increasing contracture of the thumb web.

859. Operative specimen of a Dupuytren's nodule. It is dome shaped, fibrous and attached to the palmar aspect of the palmar aponeurosis.

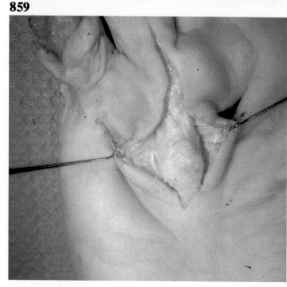

Baron Guillaume Dupuytren (1777–1835). French surgeon. Most renowned surgeon in Europe in his time. Reported palmar fascial contracture in 1833, although this was first described by Astley Paston Cooper in 1822.

860 & 861. Dupuytren's contracture with skin excoriation and intertrigo.

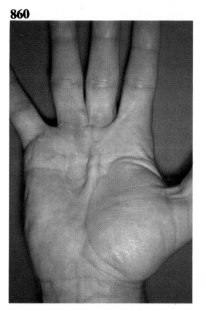

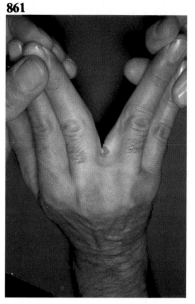

862. Dupuytren's contracture – a later stage. Flexion contracture of the proximal interphalangeal joint of the right middle finger and of all three joints of the left little finger. Contracture of the distal interphalangeal joint is rare.

863. Knuckle pads in a patient with Dupuytren's contracture. These thickenings of the dorsal fascia usually occur over the proximal interphalangeal joints but can occur over the metacarpo-phalangeal joints.

864. Dupuytren's contracture with associated plantar fasciitis (arrow).

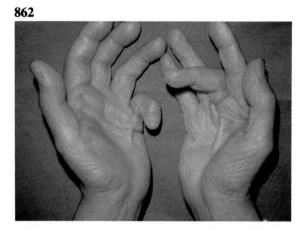

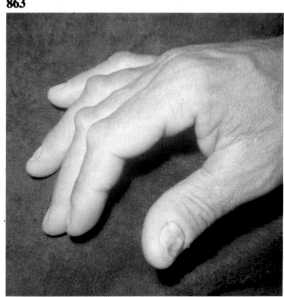

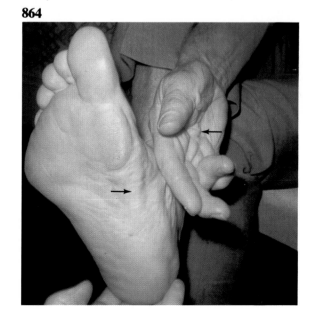

Scleroderma

Scleroderma is a collagen disease in which there is increasing fibrosis of the skin. Ischaemia and secondary trophic changes can affect the whole hand. There is a high incidence of calcinosis.

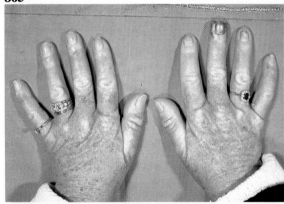

865. Scleroderma – an early case. The skin is thick and inelastic, and there is fingertip ischaemia and early atrophy of the nail. The right middle finger shows infection around the nail.

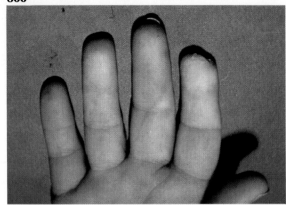

866. Scleroderma, fingertip ischaemia. Many patients have Raynaud's phenomenon.

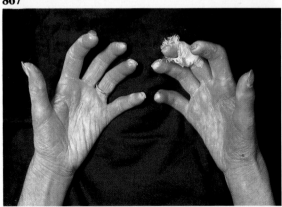

867. Scleroderma. A late stage with sclerotic skin, joint contractures, severe fingertip ischaemia, and recurrent ulcers.

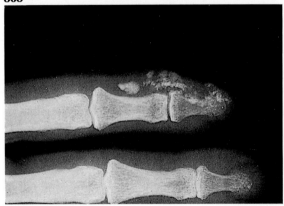

868. Scleroderma. X-ray showing calcinosis and osteoporosis of the distal phalanx.

Systemic lupus erythematosis

Systemic lupus erythematosis presents in the hand as a skin rash or arthritis.

The skin rash, which may be erythematous, papular, or purpuric, presents on the dorsum of the hand and extends along the extensor tendons to the fingernails.

The arthritic manifestations which resemble rheumatoid arthritis, can pre-date S.L.E. by years.

869 & 870. Systemic lupus erythematosis, typical skin changes.

869

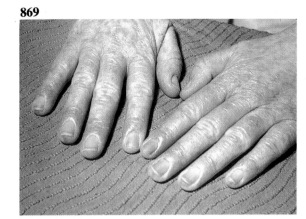

870

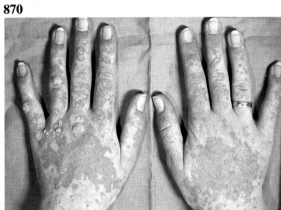

871. Polyarthritis, systemic lupus erythematosis. Swelling of the interphalangeal joints of the fingers with less obvious spindling than with rheumatoid arthritis.

871

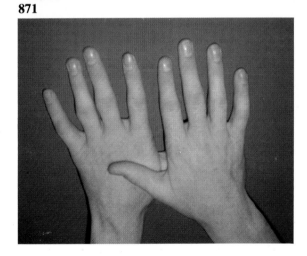

Polymyositis and dermatomyositis

In polymyositis there is degeneration and inflammation of the proximal muscles of the limb presenting as muscle weakness. Arthritis, arthralgias, and effusions of the small joints of the fingers and wrist occur in about half the patients and may be the initial manifestations of the disease.

In dermatomyositis, in addition to the above there is a dusky erythema, most marked in the hands over the metacarpo-phalangeal joints. (It does not extend along the extensor tendons as it does so often in systemic lupus erythematosis.)

50% of patients with dermatomyositis develop visceral malignancy, cancer of the alimentary canal and lungs being the most frequent. The muscular, cutaneous and arthritic features may pre-date the discovery of that malignancy by as much as two years.

872. Dermatomyositis. There is erythema over the dorsal metacarpo-phalangeal joints and the proximal interphalangeal joints. Flat topped papules can be seen over the backs of the finger joints. This is Gottron's sign and is pathognomonic of dermatomyositis.

872

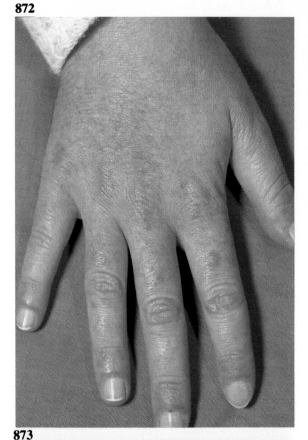

Sarcoidosis

About 50% of patients with sarcoidosis show a skin rash, and about 20% have x-ray changes in the bones of the hands. These bone changes may be punched out cystic-like areas, rarefaction or trabeculation.

Sarcoidosis in the hands may also present as peri-articular swellings resembling rheumatoid arthritis.

873. Sarcoidosis. Advanced erosions at the bases of the 4th and 5th metacarpals.

873

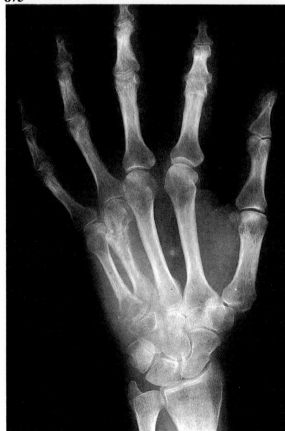

Reiter's syndrome

In this syndrome there may be arthritis of wrist and fingers, non-specific urethritis, purulent conjunctivitis, and muco-cutaneous lesions.

Hyperkeratosis and nail changes occurred in about 30% of patients. The nails become discoloured and yellow, hyperkeratotic, and are eventually shed.

874. The hands in advanced Reiter's syndrome. Note that the nail changes and the involvement of the terminal interphalangeal joints bear a close resemblance to psoriatic arthritis.

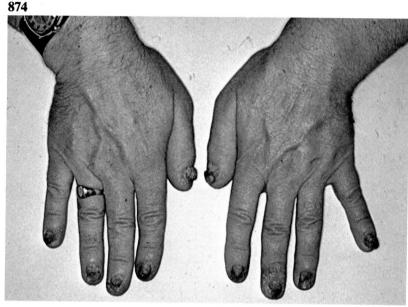

Marfan's syndrome

Marfan's syndrome is a heritable disorder of connective tissue characterised by ligament laxity. The individuals are loose, lean and long with angular elongated hyperextensive and slender limbs and digits. 90% develop cardiovascular catastrophes. Arachnodactyly, i.e. spider fingers, is associated both with Marfan's syndrome and sickle cell disease. Hyperhaemolysis and abdominal crises are not however observed in Marfan's syndrome.

875. Marfan's syndrome. X-ray showing long slender digits.

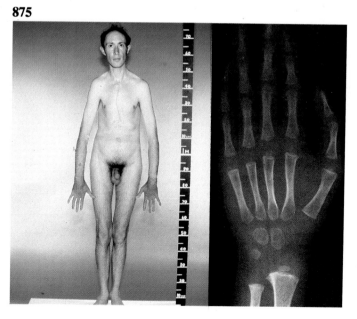

Ehlers-Danlos syndrome (The India Rubber Man)

This is a heritable disorder of connective tissue characterised by an exceptional degree of extensibility of the fingers which bend in incredibly unnatural positions, e.g. such a person can bend his fingers back until his fingernails touch the face of his wrist watch. Other abnormalities are seen in the heart, blood vessels and teeth. A bleeding tendency is manifested in these patients.

876–878. Ehlers-Danlos syndrome showing extreme extensibility of the fingers.

876

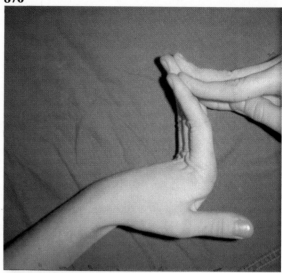

877

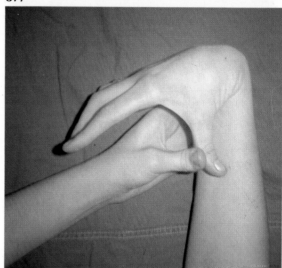

878

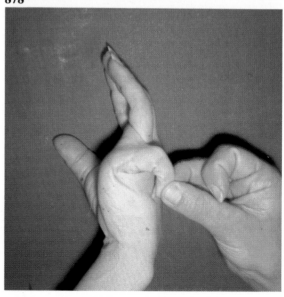

Edvard Ehlers (1863–1937). Danish dermatologist. *Henri Alexandre Danlos (1844–1912).* French dermatologist. Ehlers-Danlos syndrome. A congenital syndrome marked by hyperelasticity of the skin, rigidity of the blood vessels, excessive susceptibility of the skin to trauma.

879

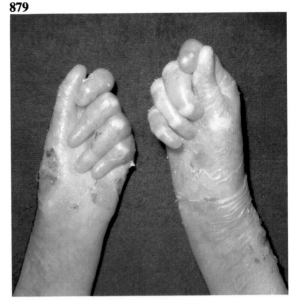

Epidermolysis bullosa

In this connective tissue disorder the elastin component is defective. The skin peels from the hand like a glove. The hand becomes clawed from acquired webbing. The fingernails undergo extreme degrees of atrophy and shortening and the nail plates become distorted, leaving deformed stubby fingertips. Usually there is associated dental deterioration and oesophageal stenosis.

879 & 880. Epidermolysis bullosa in a 22-year-old female showing peeling of the skin, clawing of the hand and atrophy of the fingertips and nails.

880

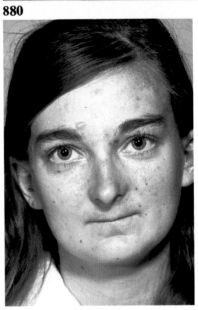

Disease of the cardiovascular and respiratory system

The hand provides many clues which help in the diagnosis of diseases of the cardiovascular system. These clues may be related to the pulse, the colour and temperature of the skin, disturbances of the nail and nail bed, and the general configuration of the hand and digits.

Blood flow determines the colour and temperature of the skin and the colour of the nail bed.

Cyanosis.
Cyanosis + warm hand = systemic disorder of blood flow.
Cyanosis + cold hand = local disorder of blood flow.

Hand signs in diseases of the cardiovascular system

CONDITION	SIGNS IN THE HAND
Congenital heart disease	Cyanosis, clubbing Polydactyly, syndactyly, arachnodactyly
Ischaemic heart disease	Shoulder/hand syndrome Palmar fasciitis Arthralgia 'Mees' lines' on the nail
Congestive cardiac failure	Veins dark and distended Nail beds cyanotic Nail moons reddish purple
Aortic incompetence	'Quincke pulse' (high pulse pressure) Capillary pulsation in the nail bed Paradoxical palmar pallor
Subacute bacterial endocarditis	'Osler's nodes', 'Janeway lesions' Petechiae Splinter haemorrhages Anaemia Clubbing
Shock	Cold, clammy, cyanotic
Raynaud's phenomenon	See page 210

Etienne-Louis Arthur Fallot (1850–1911). Professor of hygiene and legal medicine at Marseilles. His 'Contribution à l'anatomie pathologique de la maladie bleue' was published in 1888.

Tetralogy of Fallot

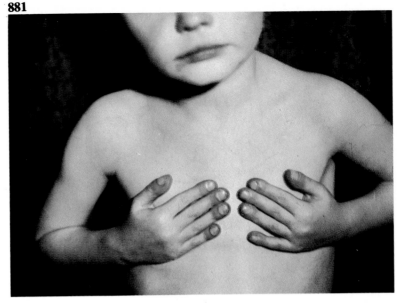

881. Tetralogy of Fallot consists of a large ventricular septal defect, pulmonary stenosis, an over-riding aorta, and hypertrophy of the right ventricle. The patient is cyanotic and has marked clubbing of the fingers.

Bacterial endocarditis

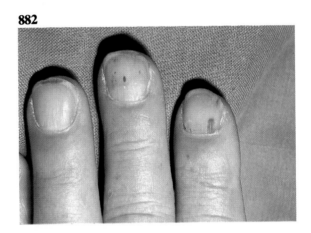

882. Bacterial endocarditis, splinter haemorrhages. Though these haemorrhages do occur in sub-acute bacterial endocarditis, the commonest cause is trauma. They also occur in rheumatoid arthritis, malignant neoplasia and in fungus infection. They are therefore a non-specific sign.

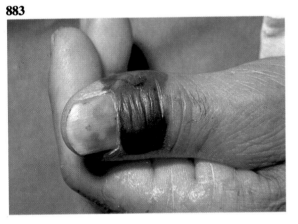

883. Acute bacterial endocarditis, due to staphylococcal infection. Spontaneous subungual and paronychial haemorrhage.

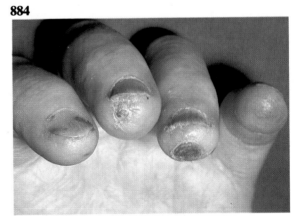

884. Congestive cardiac failure. Infarcts at the fingertips.

289

Myocardial infarction. Palmar fasciitis occurring after a myocardial infarct. See **703**.

Clubbing of the fingers

In clubbing, the fingertips become thickened and the nails are curved in both directions. There are two main types, differing in their cause and appearance.

885. Clubbing, first type. This occurs from increased blood flow through the fingertips, e.g. in congenital cyanotic cardiac disease, producing a dusky vascular clubbing. There may be a reddish brown pigmentation proximal to the lunule.

886

886. Clubbing, second type. This occurs from connective tissue overgrowth between the nail plate and bone. e.g. bronchial carcinoma, chronic pulmonary disease; this clubbing has a drier, more pallid and less engorged appearance.

Clubbing can also occur in alimentary tract disorders (cirrhosis, malabsorption states).

Clubbing is usually bilateral. However, local disturbances of circulation to one limb or one digit can cause clubbing in one hand or one digit.

885

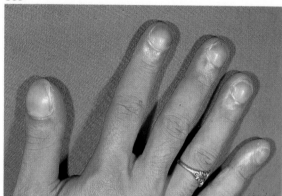

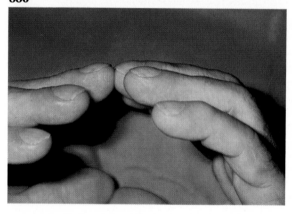

Hypertrophic pulmonary osteo-arthropathy (H.P.O.)

(Marie Bamberger syndrome)

H.P.O. is characterised by pain in the distal interphalangeal joints, and swelling of the distal phalanx with associated clubbing. There may be enlargement of the wrist.

H.P.O. may be hereditary, but it is usually associated with intra-thoracic disease, e.g. carcinoma of the lung being the commonest.

887 & 888. Pulmonary osteo-arthropathy with clubbing in a patient with carcinoma of the bronchus.

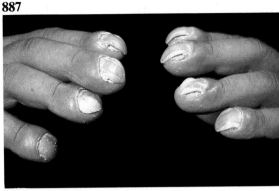

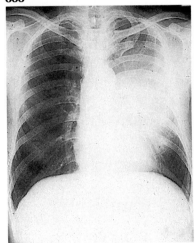

889. Associated clubbing of the toenails.

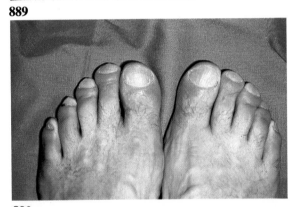

890. H.P.O. Swelling of the wrist. Periosteal bone reaction of the distal radius.

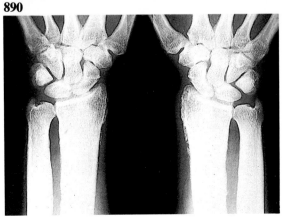

291

Disorders of metabolism

Gout

Gout presents as recurring acute mono-arthritis due to the deposition of sodium urate in the articular and peri-articular tissues. Though it usually involves the big toe (50%), the interphalangeal joints of the hand may also be involved. It can also present as soft-tissue infection.

891. Gouty involvement of the proximal interphalangeal joint. The overlying skin is thin and stretched.

892. X-ray. Clearly demarcated punched-out areas close to, but not involving, the joint.

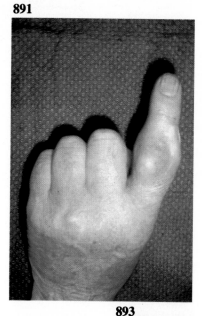

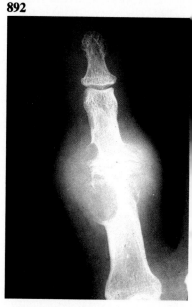

893. Chronic tophaceous gout. Some of the tophi have ulcerated and discharged a creamy white substance containing uric acid crystals.

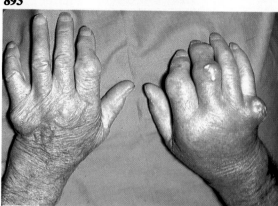

894. X-ray showing punched out areas and peri-articular enlargement.

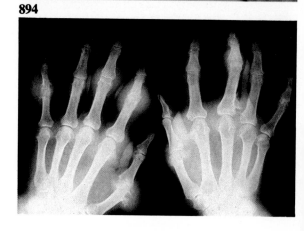

895. An ulcerating tophus.

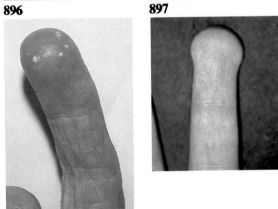

896. Gout presenting as fingertip infection (the white spots indicate gout).

897. Gout presenting as enlargement of the finger-tip.

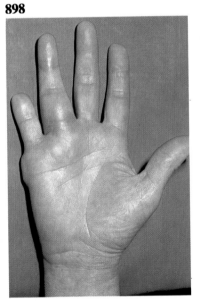
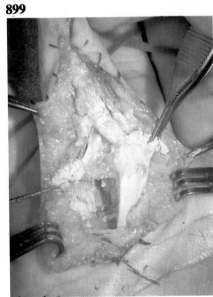
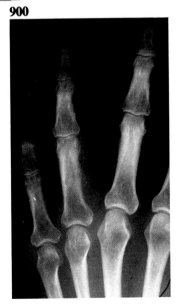

898–900. Gout presenting as an acute soft-tissue infection – right ring finger. The little finger was amputated after injury. The operative findings were chalky deposits in the tendon and surrounding tissues. The x-ray showed no skeletal abnormalities.

Hepatic cirrhosis

Patients with cirrhosis may have disorders of
pigment metabolism, hormones, coagulation and
nutrition, each of which may show as clinical signs
in the hand.

901. Hand manifestations of hepatic cirrhosis.

901

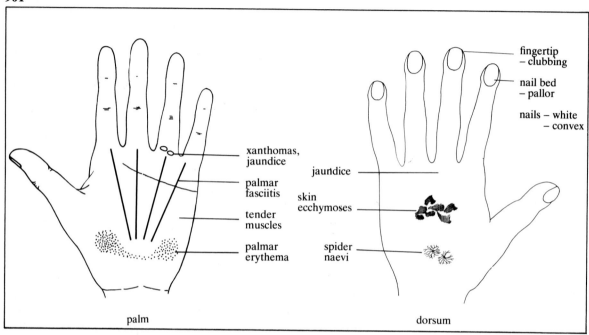

palm   dorsum

902. Obstructive jaundice due to carcinoma of the
head of the pancreas. Yellow discoloration of the
skin and sclera.

**903. Yellow discoloration of the palm compared to
the normal.**

902

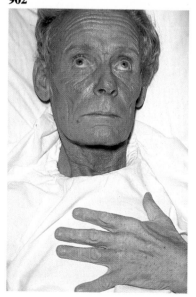

903

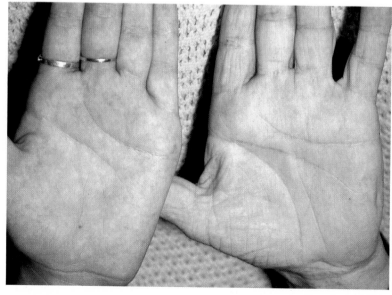

904. **An arterial spider naevus** (arrow) from hyper-oestrogenism.

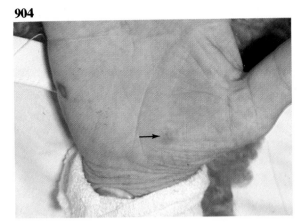

905. **Palmar erythema and prominent fascial bands** indistinguishable from early Dupuytren's contracture.

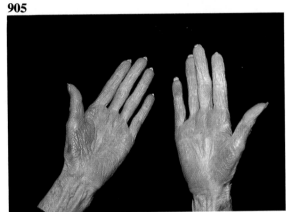

906. **Yellow discoloration** on the dorsum of the hand with ecchymoses from a coagulation defect, compared to the normal.

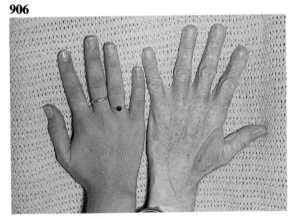

907. **Pallor of the nail bed** from associated anaemia, compared to the normal.

In jaundice the nails are often white and opaque and have a ground glass appearance.
    The fingertips may be clubbed because of an increase in blood supply.

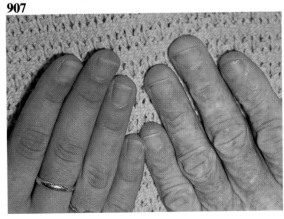

Diabetes mellitus

The hand of a diabetic may show xanthomas, evidence of diabetic peripheral neuropathy or diabetic vascular insufficiency.

908. Xanthomas are small yellow or tan papules or nodules occurring on the hand or its skin creases. They are fatty deposits occurring when the serum lipid level approximates 1,800mg/100ml.

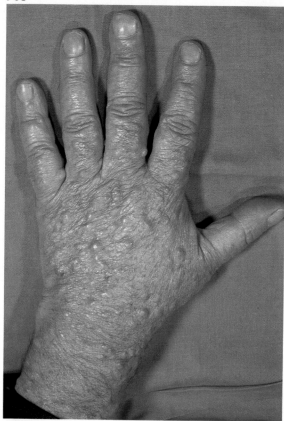

Diabetic neuropathy may manifest in the hand as wasting of the small muscles.

909. Wasting of the small muscles of the hand in a diabetic.

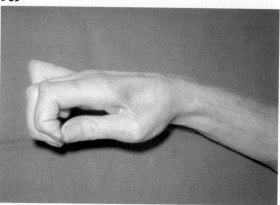

Diabetic vascular insufficiency may manifest in the hand as recurrent infections; e.g. furuncles, carbuncle, paronychia.

910. Furuncles in the hand of a diabetic. (See Chapter 4, Infections.)

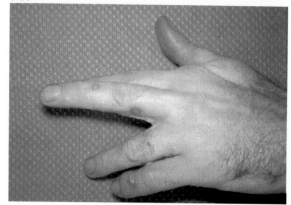

Rickets

Rickets is a disorder of calcium and phosphorus metabolism due to deficiency of vitamin D or sunshine. The earliest changes occur at the wrist and in the fingers.

911. Rickets in a young adult – characteristic enlargement of the wrist caused by symmetrical thickening of the epiphyseal plates of the distal radius.

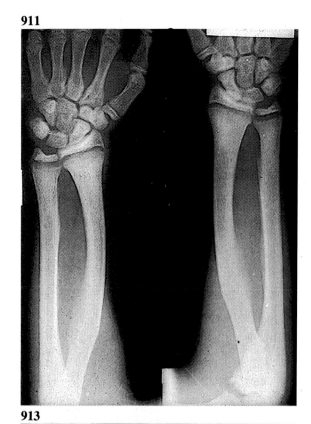

911

Porphyria

Photo-sensitivity, a prominent feature of porphyria, accounts for the characteristic skin rash seen on the hands and other exposed surfaces.

Exposure to sunlight causes blisters which rupture, leaving shallow crusted ulcerations. Secondary infections occur causing scarred deformities, especially on the tips of the fingers.

912 & 913. Porphyria in an 80-year-old woman. Her skin is rough and leathery and blistered from exposure to sun. She had red urine containing excess porphyrins.

913

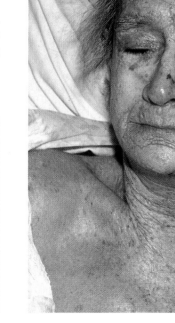

912

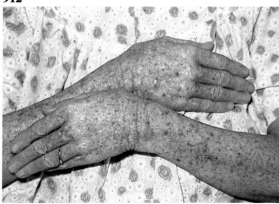

Disorders of nervous system

Many disorders of the nervous system affect the hand and its function. They may involve either the sensory or motor system alone, or both together. For autonomic disorders, refer to Reflex Dystrophy (Chapter 8).

They may be either congenital or acquired, acute or chronic, relapsing or progressive, functional or organic.

Diagnosis. In systemic disorders of the nervous system the clinical findings are diffuse, often affecting both hands and both upper and lower limbs.

When systemic nerve disorders have an insidious onset, their diagnosis is often delayed.

In peripheral nerve injury the clinical findings are confined to the distribution of one nerve.

CONGENITAL	Mongolism, Down's syndrome

ACQUIRED	
Sensory system	Syringomyelia, leprosy
Motor system	Parkinson's disease Cerebral palsy, cerebro-vascular accident, encephalitis Amyotrophic lateral sclerosis, progressive muscular atrophy Poliomyelitis Myotonia
Motor and sensory systems	Hysteria Disseminated sclerosis Peripheral neuritis (leprosy, diabetes, alcoholism) Charcot-Marie-Tooth disease

914. Sites and causes of peripheral nerve lesions.

MOTOR LOSS (WEAKNESS)

1. Anterior horn cells – anterior poliomyelitis, motor neurone disease.
2. Anterior root – arachnoiditis, neurofibroma, polyarteritis.
3. Peripheral nerve – trauma, nerve compression.
4. Motor nerve terminal – toxic neuropathy (lead).
5. Neuromuscular junction – myasthenia gravis.
6. Muscle – myotonia.

SENSORY LOSS

1. Spinal cord – multiple sclerosis, spinal cord tumour, syringomyelia.
2. Root entry zone – tabes dorsalis.
3. Posterior root – arachnoiditis, neurofibroma, tabes dorsalis.
4. Posterior root ganglion – sensory neuropathies.
5. Peripheral nerve – trauma, nerve entrapment, diabetes mellitus, alcohol.
6. Sensory nerve terminals – toxic neuropathies (arsenic).

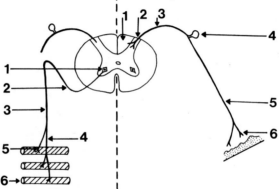

Neurological disorders of the sensory system

Syringomyelia

Syringomyelia manifests in the hand as dissociated anaesthesia – loss of heat, cold, and pain, predisposing the fingertips to trauma and infection. Light touch sensation is usually retained.

915. Syringomyelia. Trophic ulcer on the thumb. Painless and penetrating.

Later, in the course of the disease, there may be:

(a) intrinsic muscle weakness out of all proportion to the degree of wasting. A claw deformity may result. Differential diagnosis: leprosy.

(b) vasomotor changes. The hand can become cold, wet, livid, and puffy – the succulent hand of 'Marinesco'.

(c) palmar fasciitis.

915

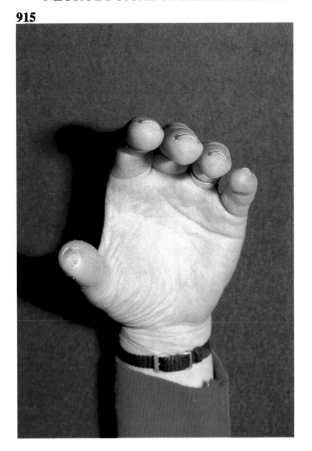

James Parkinson (1755–1824). English physician and geologist. 'An essay on the Shaking Palsy' was published in 1817. He was the first person to receive the gold medal of the Royal College of Surgeons. He also wrote political pamphlets and books on palaeontology.

Disorders of the motor system

Parkinson's disease, paralysis agitans

Parkinson's disease frequently develops asymmetrically as a slight tremor in the fingers, progressing to a pin-rolling palsy or cog-wheel phenomena. The hands at rest show a coarse and large amplitude tremor which disappears upon a conscious, purposeful movement. Emotional tension aggravates the rigidity and tremor.

916. Parkinson's disease. The hands are held rigid.

916

The spastic hand. Spastic dysfunction in the hand is seen in cerebral palsy, cerebral trauma, cerebro-vascular accidents, and after encephalitis.

917. The spastic hand in cerebral palsy. The common deformity includes pronation of the fore-arm, flexion of the wrist, the 'thumb in palm' position, flexion of the metacarpo-phalangeal joints and extension of the interphalangeal joints.

917

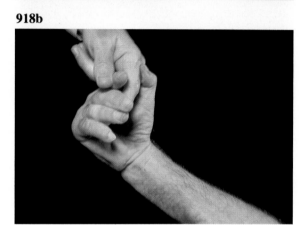

918a

918b

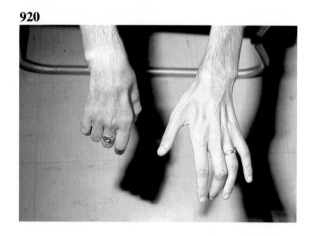

918a & 918b. Spastic right hand after cervical cord injury. This patient could not actively or passively extend his fingers or thumb.

919

919. Neurologic amyotrophy. Paralysis of right brachialis and pronator teres. See hollowing in right cubital fossa..Viral infection.

920

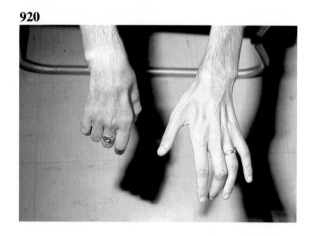

920. Amyotrophic lateral sclerosis. A relentless, incurable, motor neuron disease beginning in adult life and characterised by progressive muscular atrophy.

This patient presented first with upper motor neurone features, spastic weakness and increased reflexes, and later with severe muscle wasting. Both upper and lower limbs were affected. There were no sensory signs.

Myotonia

Muscle tone may be increased in myotonia, in extra pyramidal disorders, and in upper motor neuron defects where there is spasticity (clasp knife in character).

In myotonia the muscle, once stimulated, remains contracted for varying periods of time.

921

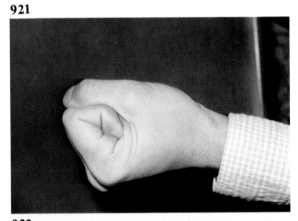

921–923. Myotonia. The sequence over four seconds when the patient was asked to squeeze the fist hard and then open the hand quickly. It opened stiffly and in slow motion. Myotonia may be noticed when shaking hands as the patient has difficulty in letting go of the grip.

0 seconds (open quickly!)

922

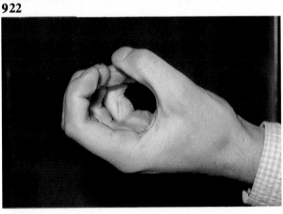

2 seconds

923

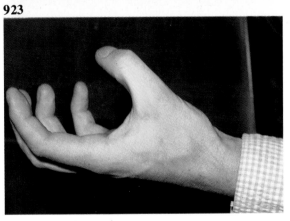

4 seconds

Disorder of both motor and sensory systems

Hysteria

Hysteria may manifest as a sensory or a motor paralysis or both. In neither case is there any anatomical basis to the clinical findings.

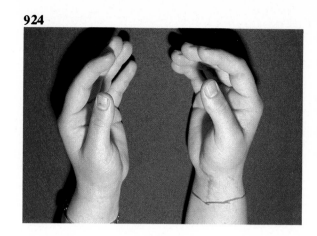

924

924. Hysteria. The sensory paralysis has a glove-type distribution affecting an entire finger or hand (see also page 202). This factory worker developed an hysterical 'glove' anaesthesia after excision of a ganglion. There were no objective abnormal physical signs.

In motor palsy the patient learns to hold his hand in certain bizarre positions. See Contractures, Chapter 6.

Multiple sclerosis is characterised by numbness and paraesthesia, often invoking the term 'a useless hand' to describe clumsiness from impairment of position sense.

Peripheral neuritis

Peripheral neuritis can be caused by a large number of factors, both local and general. There is usually a combination of both sensory and motor involvement. The common causes of peripheral neuritis include –

(a) toxicity – drugs, lead, arsenic, mercury.

(b) infections.

(c) deficiency, e.g. vitamin B complex, diabetic neuropathy, alcoholic neuritis.

(d) trauma, from direct injury to a nerve.

925

925. Toxic neuropathy, from nitro-furantoin. Involvement of the ulnar nerve is causing intrinsic muscle palsy.

Lead neuropathy. Chronic exposure to lead is followed by paresis of the upper extremities, predominantly affecting the radial nerve, leading to a wrist and finger drop. See **244**.

Diabetic neuropathy

Fifty per cent of adult diabetic patients have neuropathic symptoms and 15% have abnormal signs. They are usually distal, symmetrical, and primarily sensory, with numbness, tingling, and relatively little pain. They are usually confined to the feet and lower legs, although the hand may be affected.

926. Neuropathy in an elderly diabetic. A trophic ulcer over the base of the little finger. The index finger amputation was for gangrene.

927. Charcot-Marie-Tooth disease is a hereditary chronic degeneration of peripheral nerves. This patient presented with weakness, intrinsic muscle wasting, paraesthesia, and other slight sensory changes in both hands.

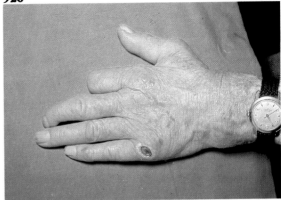

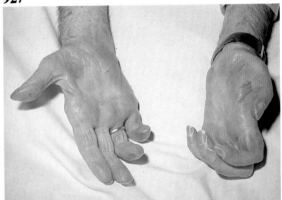

Mongolism, Down's syndrome

The hands are usually short, thick, and broad with only one transverse palmar crease. The little finger and thumb are often shorter than normal; the former is sometimes rudimentary and curves inward, toward the ring finger. There may be associated webbing of the fingers and toes and supernumerary digits.

928a. Mongolism.

928b. Hyperhidrosis.

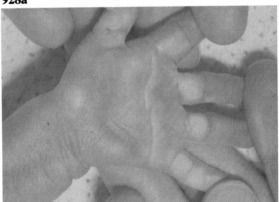

Jean Martin Charcot (1825–1893). French neurologist. The first professor of nervous diseases in the world. Charcot's disease, neuropathic arthritis, is a progressive degenerative disease of the joints, resulting from a variety of neurological disorders, including diabetic neuropathy, alcoholic neuropathy, syringomyelia, leprosy etc. The condition is constantly aggravated by the loss of pain in and hypermobility of the joint, which deprives the affected organ of natural protection from injury.

John Langdon Haydon Langdon-Down (1828–1896). Born in Cornwall. He was medical superintendent at Earlswood Asylum for Idiots in Surrey for 10 years. His article 'Observations on an ethnic classification of idiots', 1866, gave the first full description of mongolism.

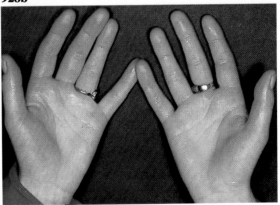

303

Disorders of the endocrine system

Acromegaly ('spade hands')

Acromegaly is caused by a pituitary tumour producing excess growth hormones. It is characterised by large spade-like hands and coarse facial features – a large nose, a big tongue, protuberant lower lip, large lower jaw, and prominent orbital ridges, nose and ears.

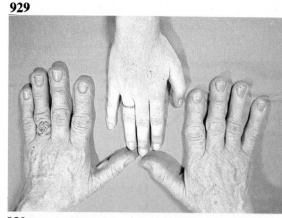

929. Acromegalic hands compared to a normal hand of the same age. The hands are elongated and broadened with coarse skin, there is an overgrowth of soft tissues, the fingers are stodgy, and the nails are broad with rectangular tips.

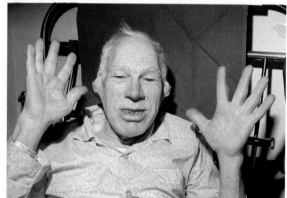

930. The acromegalic patient.

931. Acromegaly showing a spade-like palm. There is wasting of the right thenar eminence (arrow) due to median nerve compression in the carpal tunnel.

932. Large spade-like hands. There is widening of joint spaces due to hypertrophy of cartilage, tufting of terminal phalanges, and spicule osteophytes at joint margins. There may also be ossification at muscle attachments.

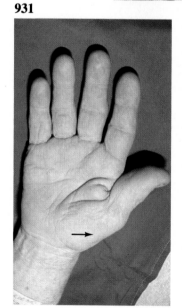

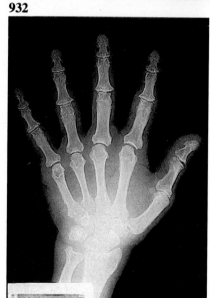

Hypothyroidism, myxoedema

933

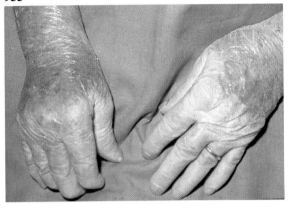

934

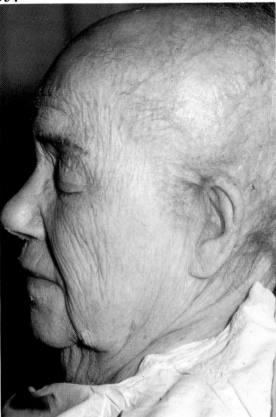

933. The hand in hypothyroidism. The hand participates in the general hypometabolic state and has a characteristic doughy, dry, rough, and stolid appearance. It usually appears larger and fatter, and the palm is pale and sallow. There may be a non-pitting type of oedema, and xanthomas are often present.

934. The face and scalp in hypothyroidism. The same woman showing the classic podgy, pallid, expressionless face, loss of hair, and thin eyebrows.

The nail changes in hypothyroidism include retarded growth, and brittleness; sometimes koilonychia, atrophy, and psoriatic pitting may be observed.

Hyperthyroidism, thyrotoxicosis

935. Thyrotoxicosis in an elderly female who presents with goitre, slight exophthalmos, and some hand changes.

936. The hands in hyperthyroidism. They are warm, wet (moist palms) and frequently show a fine tremor. The hands are often long and bony with thin fingers.

The nails may be long, shiny and curved with increased growth rate because of hypermetabolic effects. The nails may also be thickened and show vertical ridging, brittleness or loosening.

Clubbing of the fingers and isolated enlargement of the phalanges may be noted in patients with severe exophthalmos.

935

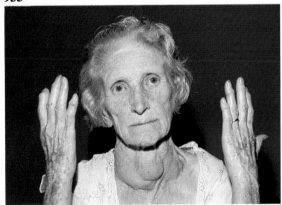

936

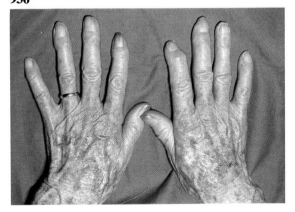

Parathyroid disorders

Parathyroid disorders, by disturbing calcium and phosphorus metabolism, affect the skin, nail and bony structure, particularly in the hand.

HYPOPARATHYROIDISM

The skin is usually dry and scaly. There may be abnormally sparse hair distribution. The fingernails may show atrophy and psoriatic-like pitting. Tetanic spasms occur when the serum calcium reaches approximately 7mg/100ml.

In tetanic spasm of the hand there is a characteristic posture in which there is bilateral flexion of the fingers of the metacarpo-phalangeal joints and extension of the interphalangeal joints. The thumbs are drawn almost to the finger, the wrists are flexed and there is ulnar deviation.

937. Carpopedal spasm due to a low serum calcium after an operation to remove a parathyroid adenoma.

938. Hypoparathyroidism. Severe pitting and atrophy of the nails and a dry scaly skin in hypoparathyroidism. The pitting of the nails is due to a secondary fungus infection.

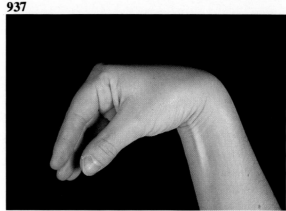

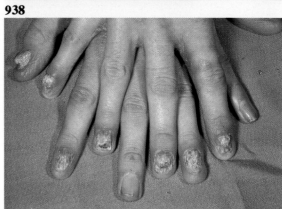

HYPERPARATHYROIDISM

Hyperparathyroidism manifests in the hands as osteoporosis, bone cysts, chondrocalcinosis and subperiosteal erosions.

The fingertips may appear bulbous and have pseudo clubbing due to the resorption and cystic changes in the distal phalanges.

939. Typical x-ray features of hyperparathyroidism. Subperiosteal resorption of the middle phalanges and of the tufts of the distal phalanges.

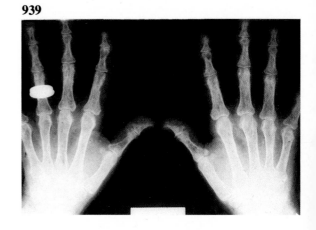

Addison's disease

Addison's disease (adrenocortical insufficiency) is associated with electrolyte imbalance and increased skin pigmentation – the latter being caused by hormonal stimulation of melanocytes which cause a generalised coppery pigment deposition in the skin. This usually appears earliest about the knuckles, on the dorsum of the hand and elsewhere on the exposed surfaces of the body. It seldom involves the palms except the creases, where it is particularly pronounced.

The fingernails often appear darker than usual.

940

940 & 941. Addison's disease in a middle-aged female. Coppery pigmentation on the skin of the face, the extensor aspect of the arms and hands.

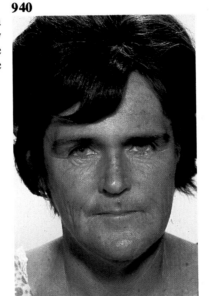

941

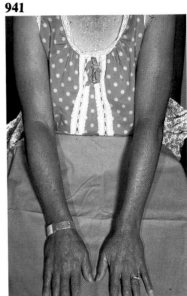

942

942. Comparison of Addison's hyperpigmentation and a normal hand of a patient of the same age.

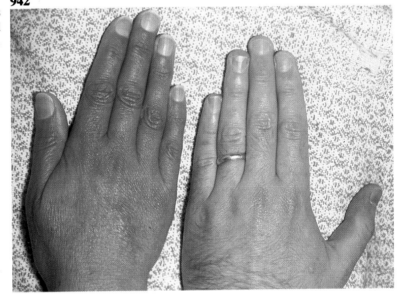

Thomas Addison (1793–1860). English physician. Described Addison's disease in 1849.

Haemopoietic diseases

Anaemia

Anaemia shows in the hand as pallor of the skin and of the nail beds. The palmar creases, normally pink, become pale when the haemoglobin concentration falls below 7g/100ml.

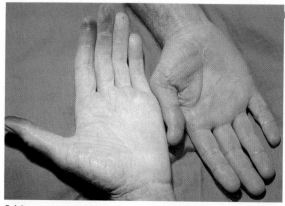

943. Anaemia. Pallor of the palmar creases contrasted with the normal.

944. Anaemia. Pallor of the nail beds contrasted with the normal. The haemoglobin was 6g%.

Nutritional defects often associated with anaemia show as nail and nail bed disorders, e.g. spoon-shaped nails, ridgings, groovings, abnormal separation of the nail from its bed.

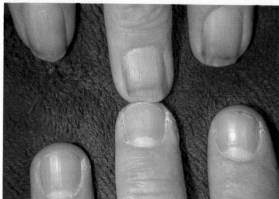

Polycythaemia

945 & 946. Polycythaemia. Pain and dusky redness of the skin of the hand and fingernails. Clubbing of the fingers is not present.

947. Gangrene of the digits from thrombosis in a patient with polycythaemia.

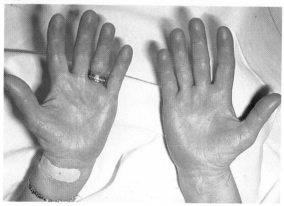

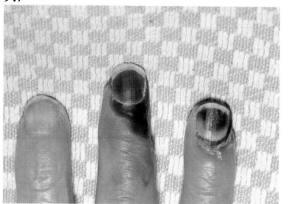

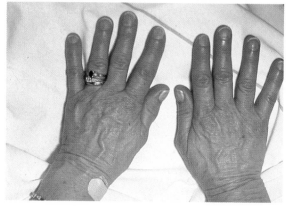

Purpura

Purpura will show in the hand because the hand is vulnerable to minor trauma and its blood vessels are precariously placed between the rigid bony structure and its thin skin.

Senile purpura shows on the dorsal skin; petechial haemorrhages are more frequently situated on the palm and pads of the fingers.

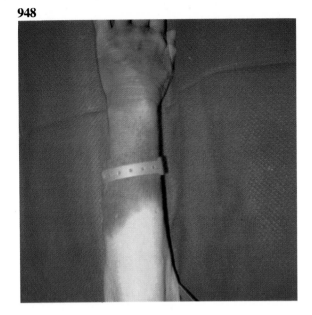

948. Purpura showing as extensive bruising in the loose dorsal skin and subcutaneous tissues.

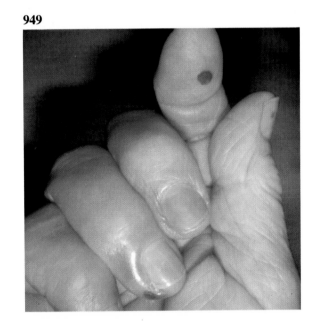

949. Petechial haemorrhages on the index and ring finger pulp. The latter became complicated by an apical pulp space infection.

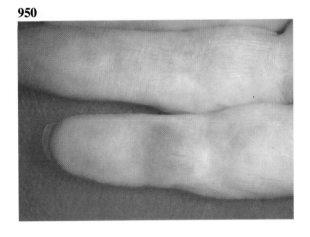

950. Spontaneous bruising on a patient taking high doses of salicylates for rheumatoid arthritis.

Miscellaneous disorders

Palmar erythema

This is a constant mottled redness in the skin beginning as a patch on the hypothenar eminence and extending to include the thenar area and the cushions in the palm overlying the metacarpal heads. Thus, it is confined to those areas of the hand which would ordinarily leave a palm print.

It occurs in hyperoestrogenic states such as hepatic disease and pregnancy, and also in rheumatoid arthritis, mitral insufficiency, vitamin deficiencies and diabetes mellitus. It can occur in healthy people.

951. Palmar erythema ('liver palms') in cirrhosis of the liver.

951

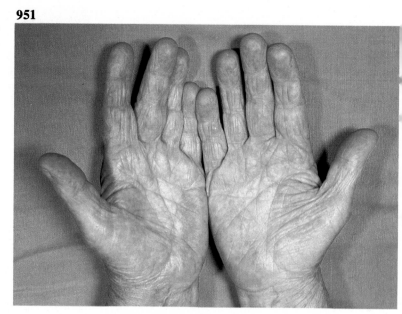

Oedema of the hand

Generalised oedema occurs in cardiac, renal, hepatic, metabolic and nutritional disorders.

Localised oedema is usually caused by a vascular or lymphatic obstruction or inflammation. See **684, 685, 689**.

Tissue fluid will occasionally accumulate in one hand or arm due to the position assumed by the recumbent patient, although the underlying condition may be a generalised oedema.

Inflammatory oedema is usually red and tender. Cyanotic oedema is usually due to venous outflow obstruction. In both these conditions there is pitting oedema. In lymphatic obstruction the oedema feels firm and indurated with absence of pitting. See **688**.

Oedema also occurs following disuse, as occurs in hemiplegia.

952

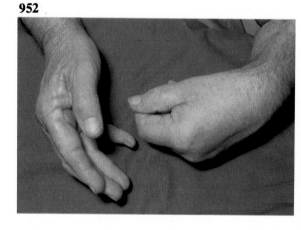

952. Disuse oedema. The left hand is swollen due to disuse oedema after hemiplegia.

Senile changes in the hand

These include a loss of manual dexterity and changes in the skin and joints.

The skin becomes drier, thinner, less resilient and more wrinkled. The veins are more prominent and senile purpuric spots, senile keratoses and palmar fasciitis are all more frequent.

There is an increased incidence of osteo-arthritis with joint deformity and Heberden's nodes.

953. The senile hand. Keratoses and purpuric spots in an 80-year-old man.

Peutz-Jeghers syndrome

In this syndrome intestinal polyposis is associated with pigmented black or dark brown spots on the palmar skin of the hands and soles of the feet, and in the mucous membrane of the mouth.

954 & 955. Peutz-Jeghers syndrome. Melanin pigmentation in the hands and mouth.

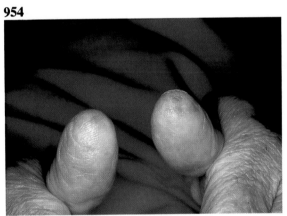

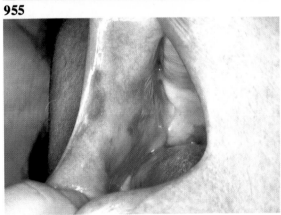

J. L. A. Peutz. Dutch. Harold Joseph Jeghers (born 1904). American physician. Peutz-Jeghers syndrome. Intestinal polyposis, cutaneous pigmentation syndrome.

Pigment changes

Pigment changes in the hand can lead to the provisional diagnosis of the following systemic diseases:

956. Pigmentation.

Brown – Addison's disease, haemochromatosis.

Yellow – pernicious anaemia, carotinaemia.

Black or brown – melanosarcoma, Hodgkin's disease.

957. Depigmentation. This occurs in: vitiligo, scleroderma, dermato-myositis, leprosy, pinta, and after dermatitis.

956

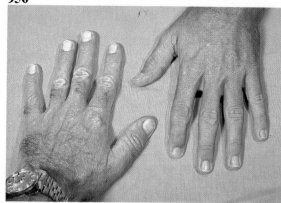

957

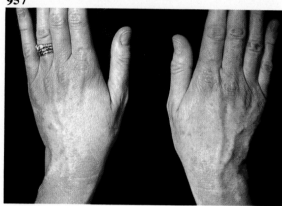

11. Nail disorders

Anatomy and function

Nails are translucent, compact plates of keratin that cover the dorsal aspect of the distal end of the digits. They form a protective covering for the fingertips and pulp. By exerting counter pressure over the fingertip they give an important function in delicate and precise touch, and in the ability to pick up tiny objects.

The nail bed is richly supplied with blood vessels, lymphatic vessels, and nerves. There are many arterio-venous anastomoses below the nail in the form of glomus bodies, which are concerned with heat regulation.

958. The normal nail apparatus.

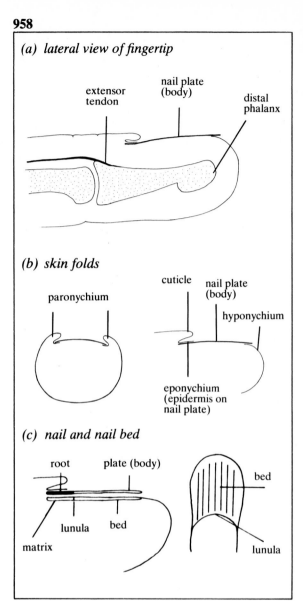

958

(a) *lateral view of fingertip*

(b) *skin folds*

(c) *nail and nail bed*

959 & 960. The normal nail apparatus, anterior and lateral view. Most of the nail has a smooth surface, but there are occasional white flecks. The angle the nail makes with the axis of the finger is less than 180°. Compare this with the clubbed finger where the nail angle with the axis of the finger is now greater than, or equal to, 180°.

959

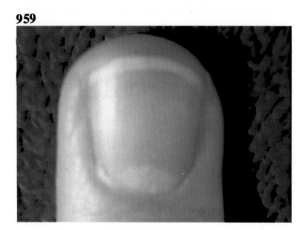

960

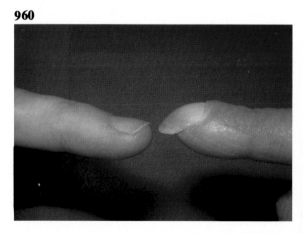

961. Subungual haematoma demonstrating the confines of the normal nail folds.

962. Subungual haematoma showing the distal edge of the nails and the extension of the haemorrhage into the lunula. The white flecks represent previous injuries or localised growth disturbances.

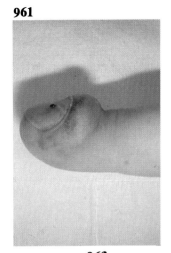

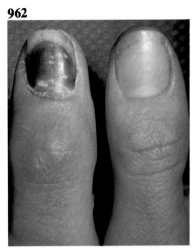

963. The nail apparatus assisting the role of pinch and in picking up a paperclip. Normal function on the right and malfunction on the left from poor nail growth.

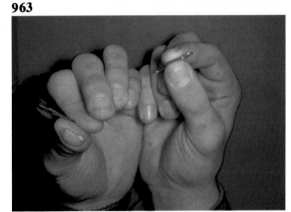

Growth

Unlike hairs, the nails grow continuously throughout life and are not normally shed. Growth is greatest in childhood and decreases slowly with ageing. The normal rate of growth of fingernails is about 1mm per week, so a nail takes about four months to reproduce itself. The nail grows from the matrix (which lies under the proximal nail fold and extends out to the nail bed as the lunula).

There are variations in growth rate. If someone is right handed, the nails of the right hand grow faster than the corresponding nails of the left, and the middle fingernails grow faster than the remainder, whilst the growth rate of the thumb and little fingers are slowest. Toenails grow at one half the rate of the fingernails.

Increased nail growth is noticed in pregnancy, warmer climates and in those who bite their nails.

Retarded nail growth occurs in malnutrition and many systemic and chronic diseases. Where there is retardation of nail growth from local or systemic nutritional defects, transverse bands may appear on the surface of the nail plate.

Alteration in structure, shape, and colour results from either local or systemic nutritional defects or injury. Localised nutritional divisions may cause grooves, ridges or transverse lines.

The nail matrix is especially sensitive to local or systemic injury or infection. It is possible to estimate the date of injury or infection by measuring the distance of a groove or ridge from the distal edge of the lunula. For example, if a transverse ridge lies 5mm from the lunula, the injury occurred about five weeks previously.

964. New nail growth after ablation three weeks previously. The new nail has grown 3mm.

965. Nail horn from incomplete surgical ablation.

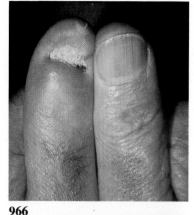

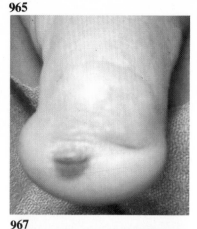

966. Bridge of an old nail remaining above a newly growing nail.

967. Beaking of the nail from lack of adequate distal pulp substance (top), and nail deformity secondary to pulp injury (bottom).

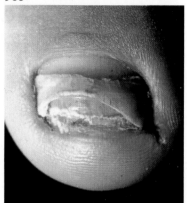

968. Severe growth disturbance from traumatic interference with the blood supply and nutrition. Note the stunting of nail growth, the generalised thickening of the nail with white flecks, and the longitudinal grooving on the right thumb nail.

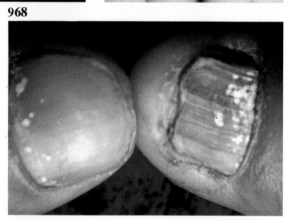

969. Nail deformity from injury to nail bed. Nail growing laterally.

Nail disorders may be primary, or secondary to other conditions, both dermatological and non-dermatological. Patients present with alterations in the shape, structure, and colour of the nail. In many instances there is no obvious cause for such disorders. See the table on page 329.

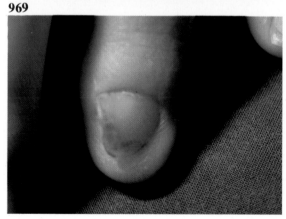

Primary conditions

Congenital, hereditary

970. Congenital absence of the fingernails. Anonychia. Left thumb and index finger; right index and middle fingers.

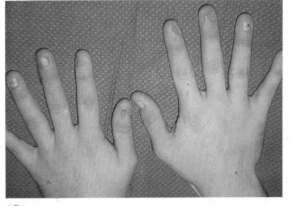

971. Hereditary bifid nail deformity of the thumb.

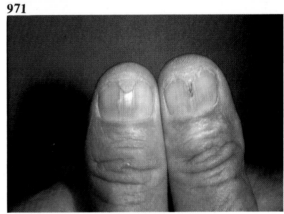

972. Micronychia. The nails are small but otherwise normal.

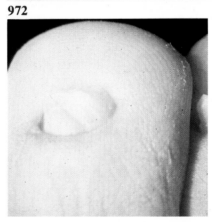

973. Senile nail changes showing diffusion of the lunula, thickening of the nail, and longitudinal ridging.

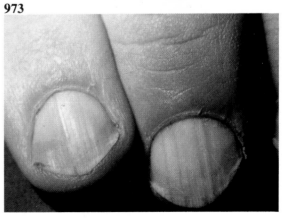

Injuries

974. Recent subungual haematoma from crush injury. This patient had severe throbbing pain which was relieved by trephining, i.e. boring a hole in the nail with release of the blood under tension.

975. An old subungual haematoma with pigmentation. See splinter haemorrhage (**1,009**).

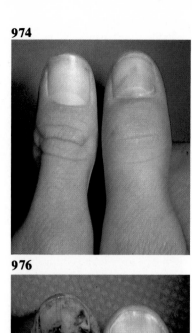

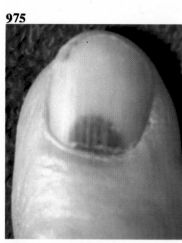

976. Traumatic dislocation of the nail plate.

977. Transverse ridges caused by manicure or tooth pressure on the matrix.

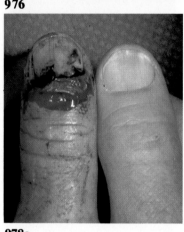

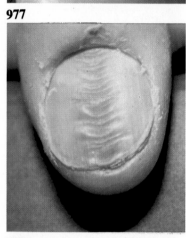

978a & 978b. Longitudinal ridging of the nail from injury to the nail bed and matrix; **978b** shows the appearance of the nail bed after removal of the nail.

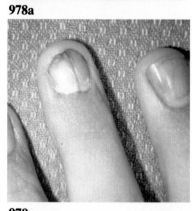

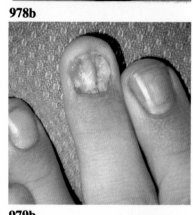

979a & 979b. Bifid nail, from injury to nail bed.

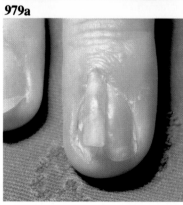

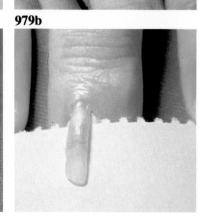

Infections

980. Acute paronychia in association with a subungual haematoma following crush injury.

981. Gram negative infection beneath the nail giving a green or bluish/black discoloration to the nail.

982. Chronic paronychia due to Candida infection. Inflammation is interfering with the development of the nail plate causing deformity.

983. Chronic paronychia with darkening and irregularity of the nail.

984. Virus paronychia. There is aggregation of vesicles around the nail folds.

985. Monilial paronychia with nail damage. There has been a thick creamy discharge from under the nail fold. The infection is causing enlargement of the nail fold and discoloration of the nail plate.

986. Fungal infection of the nail plate, typically involving the distal end. Other inflammations involve the proximal end.

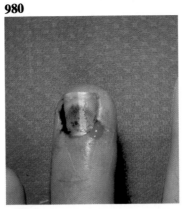

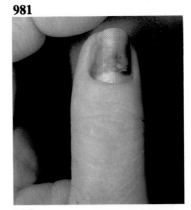

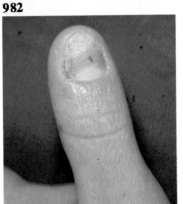

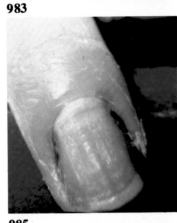

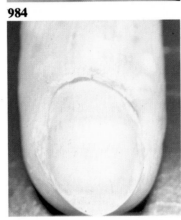

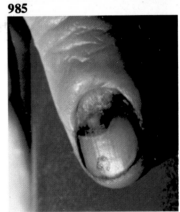

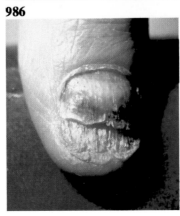

Benign tumours

987. Warts. These are the commonest tumours adjacent to the nail. They present as a roughening at the edge of the nail, which may enlarge and grow until it completely surrounds the covered part of the nail. The nail is usually unaffected, but at times the surface becomes rough and there may be cross ridges. Peri-ungual warts are often found in those who bite their nails so that the nails may show changes due to that habit.

987

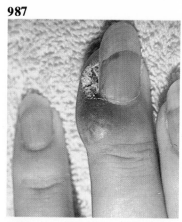

988. Mucous cyst. This is a degenerative change in the connective tissue of the distal phalanx. It forms a small cystic lesion arising from the distal interphalangeal joint and presenting between the joint and the base of the nail. If it lies over the nail matrix it causes depression of the nail.

988

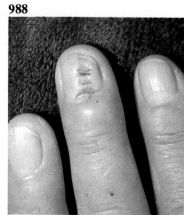

989. Giant cell tumour or enchondroma. If this tumour occurs at the distal phalanx it leads to enlargement of the fingertip, giving the appearance of finger clubbing. See **547**.

989

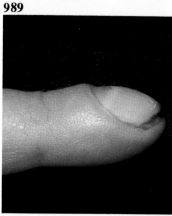

990. Pyogenic granuloma. This may occur from trauma or from association with an ingrowing nail.

990

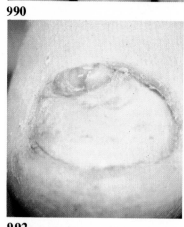

991. Glomus tumour. These may occur under the nail and present as a painful, slightly blue discoloration of part of the nail. They are exquisitely tender.

991

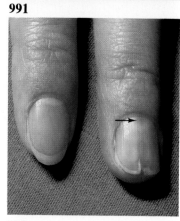

992. Fibro-adenoma causing grooving of the nail plate from pressure on the matrix.

993. Fibro-adenoma of the nail bed.

992

993

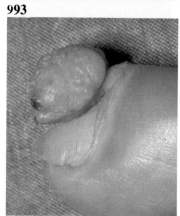

Malignant tumours

Squamous cell carcinoma of the nail bed. This is rare. It presents either as a chronic paronychia affecting a single nail and with increasing pain, or as an outgrowth from below the edge of the nail.

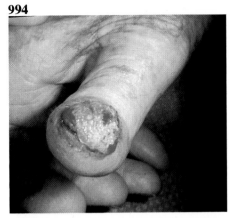

994. Squamous cell carcinoma of the nail bed. The nail has been removed.

Malignant melanoma. Subungual melanoma may present as an increasing pigmented band in the nail, a recalcitrant chronic paronychia affecting a single nail, a warty growth of the nail bed accompanied by shedding of the nail or as an amelanotic lesion, presenting as a form of pyogenic granuloma.

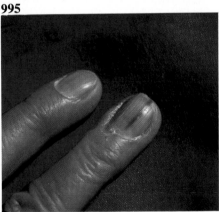

995. Subungual melanoma.

Secondary conditions affecting shape, structure, form and growth

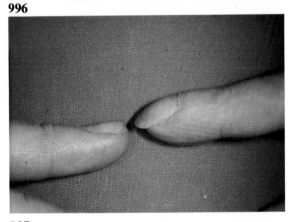

996. Clubbing of the fingernails. The earliest sign of clubbing is the loss of the obtuse angle between the nail and the dorsum of the finger which goes past 180°. The nail bed is spongy and there is marked lateral curvature of the nail, i.e. the nails are enlarged and curved in both directions.

Clubbing, also called 'hippocratic fingers', is found in chronic lung diseases (e.g. bronchiectasis), chronic heart disease (e.g. congenital heart disease or subacute bacterial endocarditis), and occasionally chronic gastro-intestinal disease. In about 10% of cases the cause if familial.

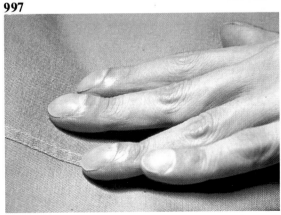

997. Clubbing of the fingers in a patient with cirrhosis of the liver. Not only are the fingernails clubbed but the ends of the fingers have a drumstick appearance that may be extreme.

Koilonychia or spoon-shaped nail. This may be congenital in origin or due to the Plummer-Vinson syndrome. Such nail deformities occur in coronary artery disease, syphilis, and in those using strong alkalis. The nails are so shaped that if a drop of water is placed on them it will not roll off.

998. Koilonychia. A case from iron deficiency anaemia.

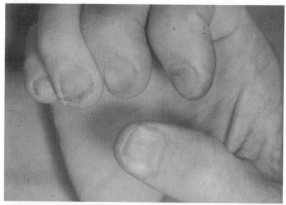

999. Early koilonychia – presenting as a brittle, flat nail.

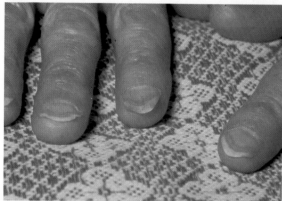

Brittleness of the nails. May be congenital or acquired. When acquired it is due to detergents, nail polish or polish removers, myxoedema, old age, or poor nutrition.

1,000. Brittle nails from poor nutrition. Though the nail looks normal it kept on breaking.

Shedding of the nails. This sometimes accompanies alopecia areata and is a minor complication of infectious fevers such as typhoid and meningitis. The nails ultimately grow again, except in severe peripheral vascular disease.

1,001. Shedding of the nails in peripheral vascular disease.

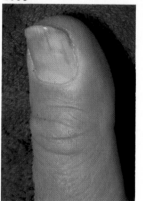

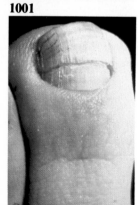

Onycholysis. Separation of the nail from the nail bed. This occurs in many conditions and is one of the features of psoriasis, fungus diseases, drug eruptions, poor circulation, Raynaud's phenomenon and trauma.

1,002. Longitudinal ridging and splitting in hyperthyroidism.

1,003. Beau's lines. Transverse ridges in the nail occur with slowing of nail growth during a debilitating illness, infectious fever or other serious upset. Growth picks up again after the illness. Local interference with blood supply can produce a similar effect.

Mees' lines. These are transverse white lines representing diminished growth. They may occur, for example, after a massive myocardial infarction.

Grooving of the nail. (1) From pressure on the germinal matrix by a mucous cyst, see **988**. (2) From pressure on the matrix by a fibro-adenoma, see **992**.

Joseph Honoré Simon Beau (1806–1865). French physician.

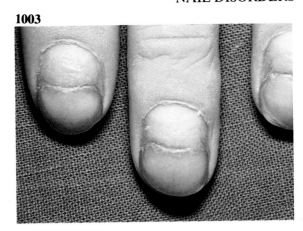
1003

Secondary conditions causing colour changes

Discoloration of the nail may be caused by:
(a) staining from external factors.
(b) degenerative changes in a nail growing very slowly.
(c) abnormal formation or partial destruction after formation of the nail.
(d) miscellaneous causes.

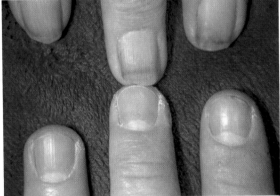
1004

1,004. Pallor of the nail bed. The nails are a good site for comparing normal and abnormal and judging the degree of anaemia present. The lower set of nails is normal. Small white flecks in the nail may be related to trauma to the nail bed when the nail was growing. The nail is translucent and shows the pinkness of the capillary bed underneath. The upper nails are also translucent but demonstrate pallor of the nail bed. The haemoglobin of this patient was 6g%.

1,005. The nails in chronic renal failure. The 'brown half moon' of chronic renal failure. Generally, the proximal half of the nail bed is pale and the distal half is red, pink or brown, but in this case it is the distal three-quarters which is brown. The normal nail is shown at the top. A white nail may be seen in chronic hypo-albuminaemia.

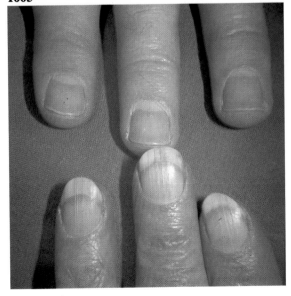
1005

Leuconychia, i.e. white spots or streaks in the nails. These may be congenital or due to trauma. They may also appear spontaneously as a result of air in the nail.

1,006. Familial leuconychia.

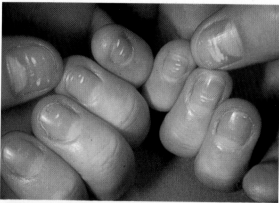

1,007. Leuconychia totalis.

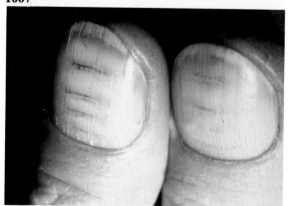

1,008. Leuconychia superficialis traumatica from crush injury and defective keratin formation.

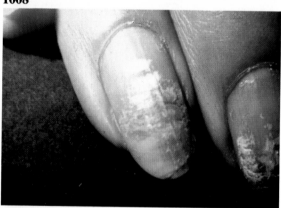

1,009. Splinter haemorrhage. These changes have by tradition been associated with subacute bacterial endocarditis, but the commonest cause is trauma. They do, however, occur in many medical conditions ranging from severe rheumatoid arthritis and malignant neoplasms to psoriasis, dermatitis, and fungus infection. They are therefore rather a non-specific sign and are of little value in diagnosis.

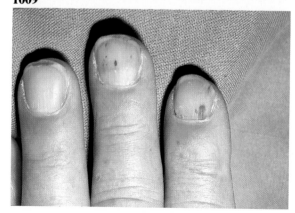

Pigmentation of the nail. This may be due to the use of potassium permanganate as an antiseptic finger bath or the taking of phenolphthalein, a constituent of some laxatives, or mepacrine, which is used in cases of chronic lupus erythematosis and in malaria.

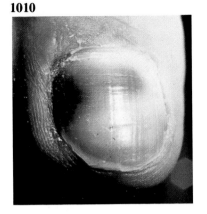

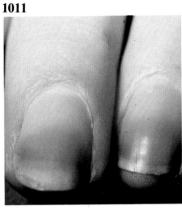

1,010. Melanonychia due to cacti injury.

1,011. Melanonychia due to nail cosmetics.

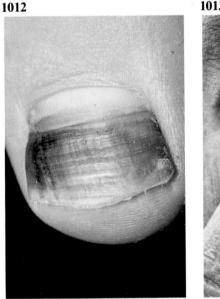

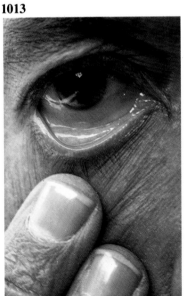

1,012. Melanonychia due to tannin.

1,013. Jaundice, carcinoma of the pancreas. In a jaundiced patient the tips of the nails sometimes have a mirror-like polish – which means he has an intractable skin irritation and may have obstructive jaundice. Also look for scratch marks and check whether the gall bladder is palpable.

Nail changes in skin disease

Psoriasis. Psoriasis is the disease which most often produces nail deformities. There are four characteristic patterns – pitting, onycholysis, discoloration, and thickening.

The earliest sign of psoriasis is pitting on the nail surface. The change may occur before the psoriasis is apparent elsewhere. Occasionally the pits are regular and form lines across the nail. Pitting may also be seen in dermatitis and chronic paronychia, and it may also occur with other diseases such as alopecia areata. The pits are due to retention of nuclei in parts of the nail keratin, which are then weaker than the surrounding normal keratin and are shed, leaving pits on the surface.

Onycholysis or separation of the nail from its bed also occurs as often as pitting. It is usually partial. There is usually a yellow margin visible between the pink and normal nail, and a separated part. Gross abnormalities of the nail plate also occur as the nail grows deformed. The thickening and yellow colour change in onycholysis is well seen in the following pictures.

1,014. Psoriatic pitting. The pits are small. In eczema large pits occur.

1014

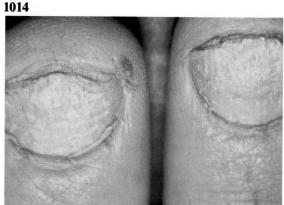

1,015. Psoriasis showing a skin rash over the interphalangeal joints and nail changes – pits, ridges, irregularity of the nail plate from the inflammatory effect of psoriasis on the matrix. Some arthropathy is evident and this may be psoriatic.

1015

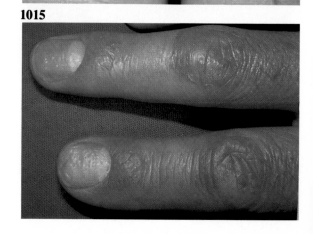

1016

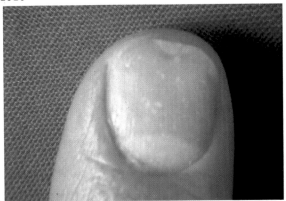

1,016. Psoriasis showing the 'oil drop' appearance; i.e. small local areas of separation of the nail plate from its bed.

1017

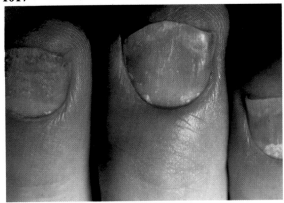

1,017. Psoriasis showing marked distal separation of the nail plate and defective keratinisation.

1018

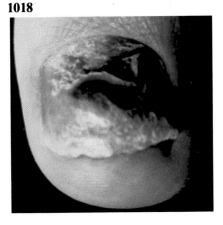

1,018. Gross changes of psoriasis which are indistinguishable from those of tinea.

1019

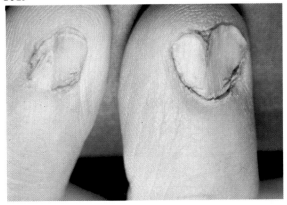

1,019. Lichen planus. Pterygium unguis and synechia. Destruction of the proximal nail plate and overgrowth of the cuticle.

1020

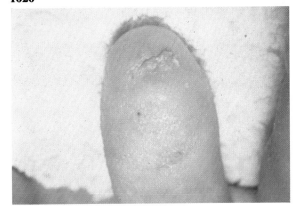

1,020. Scleroderma. Ischaemia of the nail plate causes gross dystrophy and destruction.

1021

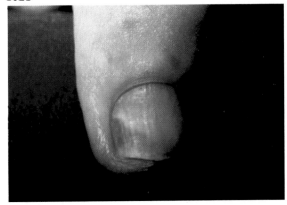

1,021. Onycholysis in Reiter's syndrome. The discoloration, ridging and pits are indistinguishable from psoriasis.

Miscellaneous disorders

General clues about the patient are provided by his or her nails. The unusual wear in the characteristic parts of the nail in watchmakers, or stains as in a photographer, may provide clues to occupation.

1,022. Indentation of the left thumbnail. This cabinet maker used a knife to trim pieces of wood held between his thumb and index finger.

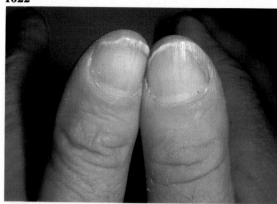

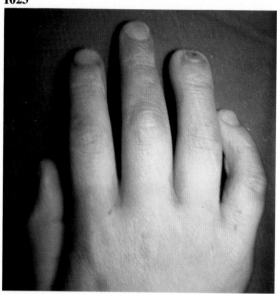

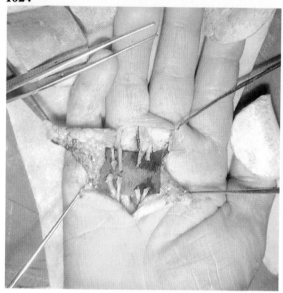

1,023 & 1,024. Nibbled finger. This child continually nibbled his right ring fingertip. Previous nerve injury had rendered it insensitive.

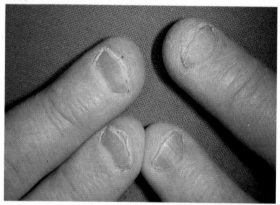

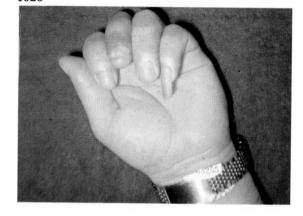

1,025. Biting of the nails may reflect an underlying emotional stress.

1,026. Long nails. Certain races keep their little fingernails long for cultural or religious reasons.

Types of nail disorder

Growth
retarded or increased
shedding

Colour
pallor
cyanosis
miscellaneous: red, green, brown, yellow, azure

Structure
grooves (Beau's lines)
ridges
lines (Mee's)
pitting
brittleness
thickness

Shape
clubbing
beaking
spoon-shaped

Miscellaneous
nail biting
congenital absence
congenital bifid nail

Causes of nail disorders

General
congenital
diet
drugs
systemic disease
dermatoses – psoriasis
psychological upset

Local
infection, e.g. paronychia
injury
poor digital circulation
occupational trauma
chemicals, e.g. detergents

Approximate incidence of nail disorders

Chronic paronychia	23%
Tinea unguium	18%
Abnormalities due to trauma	17%
Psoriasis	12%
Tumours (mucous cyst)	8%
Miscellaneous	22%

See Appendix, page 333 for a glossary of terms used in nail disease, and page 334 for differential diagnosis of nail disorders.

Appendices

Types of hand

Accoucheur's hand. This is seen characteristically in tetany, though it may occur in other spasmodic neuromuscular disorders such as athetosis, neurosis, and hysteria. There is full extension of all the fingers and the thumb at the interphalangeal joints, the four fingers are adducted firmly towards the middle finger so as to form a cone, they are semi-flexed at the metacarpophalangeal joints and the thumb is strongly adducted and opposed to the cone, of which the middle finger forms the apex.

Ape hand. The hand with a median and ulnar nerve palsy has a flattened thenar eminence and resembles the hand of a brachiating ape.

Apostolic hand. The hand with an ulnar nerve paralysis or a Dupuytren's contracture often has the ring and little fingers held in flexion.

Battledore hand (spade hand). The hand with acromegaly is usually large with a wide palm and broad stubby fingers.

Beat hand. Prolonged friction or pressure on the hand may produce cellulitis e.g. miners' beat hand.

Benediction hand. See Apostolic hand.

Claw hand. *Main-en-griffe.* The fingers are extended at the metacarpo-phalangeal joints and flexed at the interphalangeal joints from intrinsic muscle paralysis. Seen in combined median and ulnar nerve palsy at the wrist, Klumpke's paralysis, and progressive muscular atrophy etc. The claw hand without intrinsic paralysis can occur from Volkmann's ischaemic contracture of the extrinsic flexors.

Cleft hand. A congenital deformity in which there is a deep cleft in the palm, i.e. the middle finger is absent.

Crab hand. A swollen hand resulting from a scratch by the shell of a crab.

Dead hand. Those who use power hammers have repeated concussion to their hand. The effect on the blood vessels causes the hand to become dark blue and later, dead white and painful.

Drop hand. Wrist drop. Radial nerve palsy.

Flipper hand. The severe ulnar deviation of the fingers seen in chronic rheumatoid arthritis, makes the hand resemble a flipper.

Forceps hand. The hand with only a thumb and little finger, from injury or congenital deformity.

Immersion hand. After long exposure to cold and damp, the hand becomes painful and swollen.

Lobster-claw hand. See Cleft hand.

Marinesco hand. *Main succulente.* Trophic and vasomotor disorders, as seen in syringomyelia, cause the hand to become livid, cold, and wet.

Obstetrician's hand. See Accoucheur's hand.

Opera-glass hand. The gross deformity of the fingers resulting from joint dislocation in severe rheumatoid arthritis. If the distal phalanx is pulled axially, the dislocated phalanges can be extended to normal length.

Phantom hand. Following amputation the patient may feel that the fingers or hand are still present.

Preacher's hand. See Apostolic hand.

Radiologist's hand. Prolonged exposure to radiation can result in ischaemia and eventual loss of the fingers.

Skeleton hand. *Main en squelette.* The hand in progressive muscular atrophy develops gross atrophy of the muscles, making the skeleton stand out.

Simian hand. See Ape hand.

Spade hand. The hand in acromegaly. See Battledore hand.

Split hand. A congenital deformity in which the middle segments are missing and the digits on either side are fused.

Trench hand. See Immersion hand.

Trident hand. The hand in achondroplasia has short thick fingers, nearly equal in length. The trident is formed by the thumb, the index and middle fingers which are deflected to the radial side, and the ring and little fingers which are deflected to the ulnar side.

Washerwoman's hand. Hands with a white sodden-looking corrugated appearance. Also seen after drowning.

Writing hand. The hand in paralysis agitans has the tips of the thumb and index finger approximated, and the other fingers flexed.

Types of finger

Baseball finger. A flexion deformity of the distal phalanx of the finger, resulting from avulsion of the extensor tendon. It occurs during baseball and other games.

Boutonnière finger. A button-hole defect in the extensor apparatus of the proximal interphalangeal joint. It predisposes to severe finger deformity.

Bowler's thumb. A thickened interphalangeal joint or a neurofibroma on the ulnar digital nerve of the thumb can result from many years of bowling.

Clubbed fingers. The terminal phalanges are enlarged, with increase in size and curvature of the nails. The fingertips develop a parrot beak or drumstick appearance. The condition may be familial or associated with chronic suppurative pulmonary disease or due to increased vascularity of the fingertips, as can occur in congenital heart disease.

Dead fingers. Exposure to cold can cause the fingers to become white, cold, and numb.

Drumstick fingers. See Clubbed fingers.

Giant fingers. Enlargement of all tissues of the fingers is seen in macrodactyly.

Gamekeeper's thumb. Rupture of the ulnar collateral ligament of the metacarpo-phalangeal joint of the thumb occurs in gamekeepers who break the necks of rabbits and other game.

Hippocratic finger. See Clubbed fingers.

Insane finger. Paronychia is often seen in insane persons, although this is not a specific sign.

Madonna fingers. The thin delicate fingers associated with pituitary infantilism.

Mallet finger. See Baseball finger.

Murderer's thumb. The thumb which has a wide, short nail or which has a simian insertion in which the digit is located well back towards the wrist, and at more or less a right angle with the hand, invariably connotes a pathologic mentality.

Seal finger. The swollen finger seen in erysipeloid.

Snapping finger. See Trigger finger.

Spider fingers (arachnodactyly). The congenital condition in which the fingers and toes are long and slender, with excessively mobile joints.

Trigger finger (spring finger). Narrowing of the fibrous flexor sheath and/or secondary thickening of the flexor tendon, causes the finger to become trapped in flexion. Active or passive extension of the finger is accompanied by a painful jerk or click.

Tulip fingers. Dermatitis venenata due to handling tulips.

Washerman's (or washerwoman's) fingers. Immersion in water for a long time causes the fingers to shrivel.

Waxy finger. See Dead finger.

Webbed fingers (syndactylism). Adjacent fingers can be joined by skin or even skeletal tissues. A congenital deformity.

White fingers. A circulatory disease of the fingers occurring in those who use pneumatic hammers.

A glossary of terms used in congenital disorders

Adactyly. Absence of the fingers. Also called ectrodactyly (Gk *a, daktylos* finger).

Annular groove, annular band. Ring-shaped defects.

Arachnodactyly. Spider fingers; they are abnormally long (Gk *arachne* spider, *dactylos* finger).

Arthrogryposis. Congenital contracture of a joint, with articular rigidity (Gk *arthron* joint, *gryposis* curve).

Brachydactyly, brachyphalangia. All bones are present but are underdeveloped (Gk *brachys* short, *daktylos* finger).

Cleft hand, split hand, lobster-claw hand. Absence of the central fingers or rays; a form of ectrodactyly.

Clinarthrosis. Deviation in alignment of joints, usually radial and frequently bilateral (Gk *klinein* to bend, *arthrosis* a jointing).

Clinodactyly. Abnormal deflection of fingers (Gk *klinein* to bend, *daktylos* finger).

Clubhand, talipomanus. Radial or ulnar deviation of the hand, usually caused by partial or complete absence of a forearm bone (L *talipes* clubfoot, *manus* hand).

Ectrodactyly. Absence of the fingers. Partial ectrodactyly. Missing parts of fingers (phalanges or metacarpals) (Gk *ektrosis* miscarriage, *daktylos* finger).

Ectromelia. Congenital absence of one or more of the limbs (Gk *ektrosis* miscarriage, *melos* limb).

Ectrosyndactyly. Defect in which some of the digits are missing and the others are fused (Gk *ektrosis* miscarriage, *syn* together, *daktylos* finger).

Hemimelia. Absence of part of a limb (Gk *hemi* half, *melos* limb).

Hyperphalangism. The presence of supernumerary phalanges; thumb most frequently affected, with three instead of two phalanges (Gk *hyper* above, excessive).

Macrodactyly. Abnormally large fingers (Gk *makros* large).

Oligodactyly. A smaller than usual number of fingers (GK *oligo* few).

Phocomelia. Absence or impaired development of arms and forearms, but with hands present (Gk *phoke* seal, *melos* limb).

Polydactyly. Supernumerary fingers (Gk *polys* many, *daktylos* finger).

Symbrachydactyly. Condition in which fingers are short and webbed (Gk *syn* together, *brachys* short, *daktylos* finger).

Symphalangism. End-to-end fusion of phalanges of a finger, resulting in ankylosed finger joints (Gk *syn* together, *phalagx* a line of soldiers).

Syndactyly. Webbed fingers (Gk *syn* together, *daktylos* finger).

Synostosis. Congenital bony ankylosis of transversely adjacent bones, such as phalanges, radio-ulnar fusion, carpal fusion or metacarpal fusion (Gk *syn* together, *osteon* bone).

See also page 35 for a short glossary of words and prefixes used to describe congenital disorders.

A glossary of terms used in nail disease

(Gk *onyx*, L *unguis* nail)

Anonychia. Absence of nail.

Beau's lines. A transverse depression which may affect one, several or all nails (Joseph Honoré Simon Beau, b. 1806, Paris physician).

Brachyonychia. Short nail (Gk *brachys* short).

Eggshell nail. Upturning of the free border of the nail, associated with increased translucency of the nail. Seen in erythrocyanosis and Vitamin A deficiency.

Fragilitas unguium. Friable or brittle nails (L weakness).

Hepalonychia. Soft nails (Gk *hepalo* soft).

Hippocratic nails. Watch glass nails associated with drumstick fingers.

Ingrowing nail. *Unguis incarnatus.* The nail grows into the nail fold on the side causing sepsis.

Koilonychia. Spoon-shaped nails, i.e. with a concavity in the middle (Gk *koilos* hollow).

Leuconychia. White nails – partial, striate, punctate (Gk *leukos* white).

Macronychia. Large but otherwise normal nails (Gk *makros* large).

Mees' lines. Transverse white bands.

Micronychia. Small but otherwise normal nails (Gk *mikros* small).

Nail en raquette. A flattening and widening of the nail of a digit, usually of the thumb. It resembles a racquet.

Onychia. Inflammation of the nail leading to deformity of the nail plate.

Onychoatrophica. Retarded development.

Onychodystrophy. Deformity (Gk *dys-* bad, *trophe* nourishment).

Onychogryphosis. Claw deformity, i.e. when the nails curve over the tips of the fingers (Gk *gryphein* to curve).

Onycholysis. Loosening of the nail plate from its bed (Gk *lysein* to loosen).

Onychomadesis. Shedding or loss of the nail from its base (Gk *madein* to fall off).

Onychophagia. Nail biting (Gk *phagein* to eat).

Onychorrhexis. Longitudinal ridging – excessive (Gk *rhexis* cleft).

Onychoschizia. Lamination and scabbing away in thin layers (Gk *schizein* to split).

Onychosynechia. Adhesions (Gk *synechia* a continuity).

Onychotillomania. Alteration of nail structures caused by persistent neurotic picking at the nails and adjacent skin (Gk *tillein* to pluck, *mania* madness).

Pachyonychia. Thickening of the nails (Gk *pachys* thick).

Parrot-beak nail. A nail curved like a parrot's beak.

Pterygium unguis. Thinning of nail fold and spreading of cuticle over the nail plate (Gk *pterygion* wing).

Ram's horn nail. Onychogryphosis.

Reedy nail. A nail marked with furrows.

Spoon nail. Koilonychia.

Turtle-back nail. A nail which is curved both laterally and longitudinally.

Differential diagnosis of nail disorders

The condition is given in **bold**, followed by the possible disorder.

COLOUR

Pallor. Anaemia.

White marks. Injuries, arsenic poisoning, liver disease, renal disease etc.

White nails and nail beds. Hypo-albuminaemia, cirrhosis.

Yellow. Addison's disease, psoriasis, tetracyclines.

Azure. Wilson's disease.

Red (splinter haemorrhages). Injuries, subacute bacterial endocarditis, rheumatoid arthritis.

Dark red. Injury.

Green, black or blue. Gram negative infections.

Brown. Chronic renal failure, fungal and Candida infections, psoriasis.

Blue (also see **Cyanosis**). Drugs, e.g. sulphur, nitrites, mepacrine.

Cyanosis. Circulatory disorders.
+ cold hand = local cause (e.g. decreased venous return).
+ warm hand = general cause (e.g. poor pulmonary perfusion).
+ clubbing = congenital heart disease.
Note: clubbing without cyanosis = extra cardiac disease.

GROWTH

Increased. Pregnancy, warmer climate, nail biting.

Retarded. Malnutrition, systemic disease.

STRUCTURE

Transverse band, ridge, groove (Beau's lines). Impaired nutrition from local interruption of blood supply or general stress (e.g. fever, infections, surgical procedures, stress etc).

Transverse white lines (Mees' lines). Diminished growth at time of massive myocardial infarct.

Friable or brittle nails (soft splitting nails). Local trauma, constant immersion of hands in water, especially alkaline, local contact with chemicals, systemic disease such as iron deficiency anaemia, peripheral vascular disease, dietary deficiency, endocrine disorder.

Pitting. Psoriasis, alopecia areata.

Thick nails. Psoriasis, chronic trauma, peripheral neuritis, hemiplegia, peripheral vascular disease.

Loosening and lifting of the nail. Psoriasis, trauma, chemicals, hypothyroidism, systemic diseases.

Shedding of the nails. Dermatoses, systemic diseases, arsenic poisoning.

Nail biting. Neurosis.

Longitudinal ridging and splitting. Dermatoses, nail infections, senility, chemical injury, hypothyroidism, systemic diseases.

Lamination and scaling. Dermatoses, nail infections, senility, chemical injury, hypothyroidism, systemic diseases.

Thinning of the nail fold and spreading of the cuticle over the nail plate (pterygium). Vasospastic conditions (Raynaud's phenomena), hypothyroidism.

SHAPE

Spoon, concave (koilonychia). Iron deficiency anaemia, injury, diet deficiency, endocrine disorder (acromegaly, hypothyroidism).

Beaking (claw nails). Congenital, chronic system disease, traumatic loss of supporting nail bed and pulp.

Clubbing. A sign of chronic anoxia from any cause (lung disease, heart disease).

Differential diagnosis of variations in the size and shape of the hand

Spindle-shaped finger joints. Rheumatoid arthritis, systemic lupus erythematosis, psoriasis, sarcoidosis.

Cone-shaped fingers. Pituitary obesity.

Unilateral enlargement of the hand. Arterial venous aneurysm, Maffucci's syndrome.

Square, dry hands. Cretinism, myxoedema.

Single, wide, fattened distal phalanx. Sarcoidosis.

Shortened, incurved little finger. Mongolism, 'behavioural problem'.

Malposition and abduction of little finger. Turner's syndrome.

Syndactylism. In normal individuals as an inherited deformity, multiple congenital deformities, congenital malformations of the heart and great vessels.

Spider fingers (arachnodactyly). Marfan's syndrome, sickle cell disease.

Slender, delicate, hyperextensible fingers. Ehlers-Danlos syndrome, asthenic habitus, osteogenesis imperfecta, hypopituitary conditions.

Large blunt fingers (spade hand). Acromegaly, Hurler's disease.

Gross irregularity of shape and size. Paget's disease of bone, neurofibromatosis, Maffucci's syndrome.

Sausage-shaped phalanges. Rickets, tuberculous dactylitis.

Differential diagnosis of clubbing of the fingers

1. **Pulmonary causes.** Tuberculosis, pulmonary abscess, carcinoma, etc.

2. **Cardiovascular causes.** Subacute bacterial endocarditis, congenital heart disease etc.

3. **Gastro-intestinal causes.** Ulcerative colitis, dysentery.

4. **Hepatic cirrhosis.**

5. **Miscellaneous.** Familial.

Differential diagnosis of wasting of the intrinsic muscles of the hand

The intrinsic muscles of the hand are innervated by the C8, T1 nerve roots via the median and ulnar nerves.

The sites and causes of lesions which may affect the nerves or muscles and cause wasting are shown in figure **914**, page 298.

Associated clinical symptoms and signs will usually localise the site of the lesion causing the wasting.

1. Is the wasting unilateral or bilateral? If the condition is bilateral, there is probably a lesion in the neck or a peripheral neuropathy.

2. Is the wasting symmetrical or asymmetrical? If the wasting is asymmetrical, affecting either the thenar or hypothenar group of muscles, the lesion should be in either the median or ulnar nerve. If the wasting affects both muscle groups there may be a combined lesion.

3. Are there associated sensory symptoms? Wasting with sensory symptoms usually indicates a lesion of either the median or ulnar nerve or both. Wasting without sensory symptoms usually indicates a spinal cord tumour, syringomyelia or spondylitis, affecting the first thoracic nerve root.

4. Are there other signs in the hand itself? Trophic ulcers occur in syringomyelia, fasciculations in motor neuron disease.

5. Are there signs in the eyes? If a Horner's syndrome is present the lesion is in the lower cord of the brachial plexus (C8, T1).

6. Are there signs in the face? Thenar wasting from median nerve compression may be associated with myxoedema or acromegaly.

7. Are there signs in the feet? A foot drop may indicate motor neuron disease or polyneuritis. Pes cavus may indicate peroneal muscular atrophy.

Differential diagnosis of weakness of the hand (e.g. weak pinch)

1. Diminished muscle power, e.g. nerve palsy, myotonia, or muscle inco-ordination.

2. Impaired tendon glide, e.g. frictional teno-synovitis.

3. Joint disorder, e.g. arthritis.

4. Sensory nerve loss, both cutaneous and postural, e.g. nerve injury, syringomyelia.

5. Miscellaneous, e.g. functional or hysterical conditions.

Differential diagnosis of gangrene in the hand

Definition. Macroscopic necrosis of tissue caused by the blocking of its blood supply.

The clinical features depend on the repetitive onset of the ischaemia and the amount of fluid contained in the tissues.

Dry gangrene results from arterial ischaemia in desiccated tissue. Bacterial growth is minimal. Black tissue demarcates from viable tissue.

Moist gangrene occurs when there is venous as well as arterial obstruction. The tissue becomes swollen, soggy, and infected. This type of gangrene spreads rather than demarcates.

Causes of gangrene.

1. Ischaemia.
(a) Thrombosis (atherosclerosis etc.).
(b) Embolism.
(c) Diabetic gangrene.
(d) Raynaud's disease and phenomenon (cervical rib).

2. Infective.
(a) Acute inflammatory gangrene.
(b) Gas gangrene. e.g. clostridial infection. Note that gas is sometimes seen on x-rays after an open fracture.

3. Trauma.
(a) Direct – from direct mechanical injury.
(b) Indirect – where injury causes necrosis at a distance from the injury, by damage to main blood vessels.

4. Thermal.
(a) Burns.
(b) Frost bite.

Differential diagnosis of oedema of the hand

1. Bilateral oedema of the hand.
Cardiac, renal, hepatic failure.
Superior vena caval obstruction.
Hypo-proteinaemia.

2. Unilateral oedema of the hand.
Venous obstruction.
Lymphatic obstruction.
Congenital with or without constriction rings.
Secondary to axillary pathology (lymphoma).
Injury.
Infection.
Allergy.
Artefacta.

Pain in the hand – differential diagnosis between nerve root and more peripheral nerve disorders

Nerve root disorders

CONDITION	HISTORY	EXAMINATION	COMMENT
Cervical spondylosis	Sharp pains radiate to supraclavicular area and side of neck and down the limb; aggravated by neck movements	Tenderness over posterior neck muscles Restricted cervical spine movements Exacerbation of symptoms by lateral cervical flexion towards the affected side Lower motor neurone signs in cervical nerve root areas	Abnormal clinical signs can occur both proximal and distal to the wrist X-ray features of cervical spondylosis occur in most people over 40 years
Thoracic outlet (cervical rib) syndrome	Pain radiates from the neck down the inner side of upper limb to the hand Associated vasomotor symptoms	Sensory and motor signs of C8, T1 nerve roots, ulnar nerve or medial cutaneous nerve of forearm Diminished wrist pulses (on elevation of the upper limb and turning the head to the opposite side) Exacerbation of symptoms by lateral cervical flexion away from the affected side	Vascular signs in the neck X-ray may show cervical rib *Differential diagnosis:* Pancoast's tumour

Peripheral nerve disorders

CONDITION	HISTORY	EXAMINATION	COMMENT
Peripheral neuritis	Acute onset of pain, numbness, and weakness, especially in an alcoholic or diabetic patient	Sensory signs – diminished sensation Motor signs – weakness, palsy Diminished reflexes The involved superficial nerves may become palpable	Mostly bilateral May affect all limbs
Carpal tunnel syndrome	Pain and paraesthesia (in median nerve distribution) Radiates proximally Nocturnal 'waking numbness' Mostly idiopathic, but can be associated with hormone change e.g. pregnancy, menopause, or flexor tenosynovitis e.g. rheumatic	There may be no abnormal signs Hypo- or hyper-aesthesia in median nerve distribution (excluding palmar cutaneous branch) Phalen's wrist flexion sign Tinel's tapping sign Weak abductor pollicis brevis and opponens pollicis	No abnormal signs proximal to the wrist, but symptoms may radiate to the shoulder Cervical spondylitis may co-exist but the condition rarely causes nocturnal pain
Median nerve compression in forearm, 'Pronator syndrome'	Repetitive heavy work involving forearm and hand	Median nerve signs with involvement of palmar cutaneous branch, ulnar nerve signs with involvement of dorsal branch Tinel's sign at forearm, pain on resisted pronator teres action, negative Phalen's wrist flexion sign	
Ulnar nerve compression at elbow	Elbow disorder. Osteo-arthritis. Cubitus valgus Trauma or occupational Post-anaesthetic complication Paraesthesia, numbness or weakness in ulnar distribution of hand	Positive Tinel's sign over ulnar nerve at the cubital tunnel Thickened ulnar nerve Associated medial epicondylitis	May require nerve conduction study to help in diagnosis
Ulnar nerve compression at wrist	May follow trauma Weakness with a little numbness	Sparing of dorsal branch of ulnar nerve May be pure motor involvement	X-ray for fracture of hook of hamate Exclude ganglion at piso-hamate articulation *Differential diagnosis:* pisohamate osteo-arthritis

Differential diagnosis of ulcers of the finger or hand

Definition. An ulcer is a localised necrotic lesion of the skin in which the surface epithelium is destroyed and deeper tissues are exposed in an open sore.

Ulcers of the finger or hand may be due to:

Ischaemia. Poor digital circulation from atherosclerosis, diabetic vascular disease, Raynaud's disease, sclerodactyly, post-irradiation, endarteritis etc.

Infection. Acute pyogenic. Specific bacterial disease – tuberculosis, syphilis, anthrax. Fungal intertrigo.

Neoplasia. From pressure of a malignant growth or due to replacement of the skin by a new growth, which then breaks down by ulceration e.g. basal cell carcinoma, squamous cell carcinoma.

Neurogenic lesions. Peripheral nerve injury. Leprosy. Syringomyelia.

Trauma. *(a) Mechanical.* Usually repetitive trauma, e.g. self-inflicted wounds, iatrogenic pressure sore.

(b) Burns. Chemical, thermal, electrical, irradiation, frost bite.

Differential diagnosis of crepitus in the hand

Definition. Crepitus is the term most often used to mean the grating or cracking sensation or noise when two rough surfaces are being rubbed together.

Causes of crepitus in the finger or hand.

Bone. Fracture. Tumour (secondary carcinoma, myeloma).

Joint. Osteo-arthritis. Osteophytes. Loose bodies.

Tendon. Frictional tenosynovitis (de Quervain's syndrome). Trigger finger.

Subcutaneous and fascial planes. Clostridial infection. High pressure injection injuries.

Diagnosing the cause of a joint contracture by altering the posture of the more proximal joint (see also page 186)

CAUSE OF CONTRACTURE

1. Extrinsic muscle fibrosis (e.g. Volkmann's contracture)

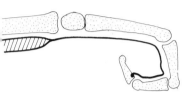

claw hand

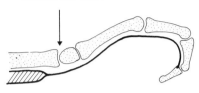

extend the wrist

the flexors tighten and prevent passive extension of fingers

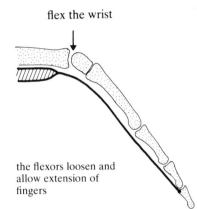

flex the wrist

the flexors loosen and allow extension of fingers

2. Intrinsic muscle fibrosis

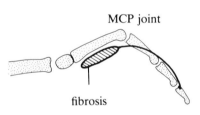

MCP joint

fibrosis

fingers stiff in extension at IP joints

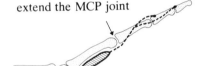

extend the MCP joint

the intrinsics tighten and prevent flexion of IP joints

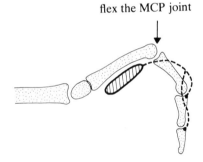

flex the MCP joint

the intrinsics loosen and allow passive flexion of IP joints

3. Flexor tendon adhesions (causing flexion contracture of the PIP joint)

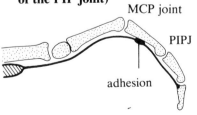

MCP joint

PIPJ

adhesion

finger stiff in flexion

adhesions distal to MCP joint

PIP joint

A

B

flexion of MCP joint has no effect on distance A–B and does not allow extension of PIP joint

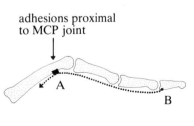

adhesions proximal to MCP joint

A

B

flexion of MCP joint reduces distance A–B and allows extension of PIP joint

INDEX

Each chapter has its own list of contents in which every picture is mentioned, and the reader should refer to these for the rapid location of particular conditions. They are on the following pages: the normal hand 11, congenital disorders 35, injuries 58, infections 120, tumours 144–145, contractures and deformities 181, vascular disorders 204, painful disorders 220, occupational conditions 252, systemic disease 272, nail disorders 313.

In the index, numbers in light type refer to page numbers, those in **bold** to picture and caption numbers.

INDEX